Andrea Zinzani

# The Logics of Water Policies in Central Asia

Andrea Zinzani

# The Logics of Water Policies in Central Asia

## The IWRM Implementation in Uzbekistan and Kazakhstan

LIT

The publication of this book was possible thanks to the support of the Conseil Universitaire of the University of Fribourg/Freiburg (CH).

This book is printed on acid-free paper.

**Bibliographic information published by the Deutsche Nationalbibliothek**
The Deutsche Nationalbibliothek lists this publication in the Deutsche Nationalbibliografie; detailed bibliographic data are available in the Internet at http://dnb.d-nb.de.

ISBN 978-3-643-90645-8
Zugl.: Verona and Fribourg/CH, Univ., Diss., 2014

**A catalogue record for this book is available from the British Library**

© LIT VERLAG GmbH & Co. KG Wien,
Zweigniederlassung Zürich 2015
Klosbachstr. 107
CH-8032 Zürich
Tel. +41 (0) 44-251 75 05    Fax +41 (0) 44-251 75 06
E-Mail: zuerich@lit-verlag.ch    http://www.lit-verlag.ch
**Distribution:**
In the UK: Global Book Marketing, e-mail: mo@centralbooks.com
In North America: International Specialized Book Services, e-mail: orders@isbs.com
In Germany: LIT Verlag Fresnostr. 2, D-48159 Münster
Tel. +49 (0) 2 51-620 32 22, Fax +49 (0) 2 51-922 60 99, E-mail: vertrieb@lit-verlag.de
In Austria: Medienlogistik Pichler-ÖBZ, e-mail: mlo@medien-logistik.at
e-books are available at www.litwebshop.de

# ACKNOWLEDGMENTS

This book is the result of more than three years of doctoral studies from January 2011 to March 2014; I can state with emphasis that a PhD is a long and intense experience of life according to different perspectives: it is a learning process, based on an everyday grow up, an increase of knowledge, culture, experiences and social exchanges. An intimate, deep and wonderful process which I think allow you an in-depth understanding of the complex social mechanisms and processes that underlie the world.

This learning process and finally this book was possible thanks to the help and support of many individuals to whom I would like to express my gratitude: to my thesis supervisors Prof. Bichsel and Prof. Salgaro for their excellent guidance of my thesis and for the freedom they gave me in structuring my research. I am grateful to Prof. Faggi, the first person to whom I explained my research idea, and helped me to build up a proposal. Since the first day until now he has always supported my research, and his long experience strongly influenced my reflections on the relations between political power, water management and territories.

I owe gratitude also to Prof. Olivier Graefe and to Dr. Olivier Ejderyan for their support and help during my stay at the University of Fribourg. Thanks so much to the all the geographers of the Geography Unit in Fribourg for the great social environment, the different discussions and the nice time spent together.

I am strongly grateful to the Italian geographers, Dr. Francois Bogliacino, Dr. Matteo Proto, Dr. Emanuele Frixa, and Dr. Federico Ferretti for our discussions, reflections and the nice time spent together in these years, in particular in the "ufficio" starting from 6 P.M.

I am still indebted also to the members of the Archeological Expedition of the University of Bologna, Dr. Mantellini, Dr. Rondelli, Dr. Bonora and Prof. Tosi, who allowed me to discover Central Asia in 2005, and gave me the basis on how to structure and conduct a field-research.

A special thanks also to all the people who helped and supported me during my field-work in Uzbekistan and Kazakhstan: to the staff of the International Water Management Institute (IWMI) in Tashkent, for the reflections and the support to my research; to my research assistants Ravshan Pirnazarov, Begzod Khuramov, Fazilat Radjabova, Dima Kostiushkin and the other friends in Samarkand, and to

Dina Baidildayeva, Nazym Dussupova and Serghei in Shymkent for their help in logistic, establishing contacts and translating interviews; to all the members of the Academy of Sciences, experts, state-province and district water bureaucrats, water users, farmers and the villages' drunk fools of the Zeravshan and Arys valleys for allowing me the understanding of their complex and conflicting socio-political relations with their hydraulic territories.

I am most grateful to my family for their enduring support, encouragement, care and interest in my research.

Finally thanks to Alice, who has shared with me this adventure for a long time and has been patience during my long stays abroad in these years; her love and care enabled me to complete this path.

# CONTENTS

## 2. COMPARATIVE METHODS IN WATER STUDIES: THE METHODOLOGICAL APPROACH

## 3. WATER POLICIES IN THE CENTRAL ASIAN REGION: FROM THE HYDRAULIC MISSION TO THE IWRM

# 6. THE LOGICS OF THE BASIN / LOCAL LEVEL WATER REFORMS IN KAZAKHSTAN: THE ARYS VALLEY

## 6.1 A GEOGRAPHICAL OVERVIEW OF THE VALLEY

## 6.2 MANAGING WATER AT THE BASIN LEVEL: DISPUTES OF SCALE AND AUTHORITIES

# GLOSSARY

The transliteration of Russian words follows the Library of Congress system according to the following table. In the text, words in italic denominate terms in Russian, Uzbek or Kazakh. Plural form of Russian words are rendered with a simple transliteration of the Russian plural. Here follows the translation of the used terms.

Library of Congress Transliteration

| Cyrillic | Transliteration |
|---|---|
| А а | a |
| Б б | b |
| В в | v |
| Г г | g |
| Д д | d |
| Е е | e |
| Ж ж | zh |
| З з | z |
| И и | i |
| Й й | ï |
| К к | k |
| Л л | l |
| М м | m |
| Н н | n |
| О о | o |
| П п | p |
| Р р | r |
| С с | s |
| Т т | t |
| У у | u |
| Ф ф | f |
| Х х | kh |
| Ц ц | ts |
| Ч ч | ch |
| Ш ш | sh |
| Щ щ | shch |
| Ъ ъ | " |
| Ы ы | y |
| Ь ь | ' |
| Э э | e̲ |
| Ю ю | i̲u̲ |
| Я я | ia |

**aryk** (Uzbek): small water canal
**akimyat** *(Kazakh):* municipality
**fermer** *(Russian)*: farmer
**kolkhoz** *(Russian)*: collective farm

**kommunalnivodkhoz** *(Russian*, used in Kazakhstan) district water department
**limit** *(Russian)* water quota
**minvodkhoz** *(Russian)* ministry of water resources
**miraab** *(Uzbek)*: person who allocates water at the local/farm level
**oblast** *(Russian)*: province
**oblastvodkhoz** *(Russian)*: province water department
**rayon** *(Russian)*: district
**rayonvodkhoz** *(Russian,* used in Uzbekistan): district water department
**RGP** *(Russian,* used in Kazakhstan): republican state enterprise
**Sai** *(Uzbek-Kazakh)*: stream, small river
**Selkhoz** *(Russian)* agricultural department
**shirkat** *(Uzbek)*: joint stock company
**sovkhoz** *(Russian)*: state farm
**SIU***(Uzbek)*: water users association
**SPKV** *(Russian,* used in Kazakhstan) water users association
**Tomorka** *(Uzbek)* household garden
**tugai** *(Russian)* riparian forest located in fluvial areas in arid environements
**vodkhoz** *(Russian)* water department

# ABBREVIATIONS

---

ADB: Asian Development Bank
BISA: Basin Irrigation System Agency
BWO: Basin Water Organization
CWR: Committee of Water Resources
DWD: District Water Department
GIS: Geographic Information System
GIZ: German International Cooperation
GWP: Global Water Partnership
ICAS: Interstate Council on the Aral Sea
ICWC: Interstate Commission for Water Coordination
IMT: Irrigation Management Transfer
ISA: Irrigation System Authority
ISF: Irrigation Service Fee
IWRM: Integrated Water Resources Management
NGO: Non-Governmental Organizations
O&M: Operation and management
RBC: River Basin Council
SANIIRI: Central Asia Research Institute for Irrigation
SDC: Swiss Development Cooperation
SSR: Soviet Socialist Republic
UN: United Nations
USAID: United State Agency for International Development
USSR: Soviet Union
WB: World Bank
WUA: Water Users Association

# ABSTRACT

---

Since ancient times water resources have always played a strategic role in societal and territorial development. These dynamics have been even more strategic in the arid and semi-arid regions throughout the world where rivers were diverted and canal systems were designed in order to increase the irrigated lands and to allow the development of societies, as occurred for instance along the Nile, Indo, Tigri and Eufrate valleys. In the past the relationship between water control and the emergence of strong despotic states led to the rise of the so-called *hydraulic societies,* based on the experiences arising within agricultural societies characterized by state-centralized water works.

Whereas until the 1970s water resources management was considered a technical issue mostly under the control of state-centralized water bureaucracies, in the last decades a significant change has occurred due to different issues ranging from population growth and relevant political changes to land degradation and inequitable water access. Therefore it emerged that the management of water resources is not merely a technical issue, but a sociopolitical, economic, and environmental one which involves a wide spectrum of different actors throughout the society and their rationales. Nevertheless in most of the developing countries these changes were influenced and somehow hampered by political and economic issues.

In order to strengthen these processes since the 1990s several international donors and agencies have sought to promote a new water management framework, the Integrated Water Resources Management (IWRM), based on the "Dublin Principles", discussed at the conference on "Water and the Environment" 1992. The IWRM –and the implementation of its pillars- aims to improve water resources management according to a multi-perspective sustainability: environmental, economical, social and political ones. Moreover, to strengthen this framework and its rationale, the Irrigation Management Transfer (IMT) and the creation of the Water Users Associations (WUAs) were supported. Although guidelines to support the implementation in the different states were provided, in the last decade a wide debate among members of the academia and water professionals have emerged discussing its definition, its pillars and in particular the implementation procedures throughout the world.

The present research aims to analyze and understand the *Integrated Water Resources Management* (*IWRM*) implementation processes, its logic and related water issues, in post-Soviet Central Asia—a wide heterogeneous arid and semi-arid region mostly included in the Aral Sea basin. These processes will be analysed in Uzbekistan and Kazakhstan, focusing at the local level (Middle Zeravshan valley –UZB-, Arys valley –KAZ-) according to a comparative approach. The following questions arise: What is the logic which has affected the IWRM implementation? Were the national sociopolitical systems able to shape this process according to their strategies and aims? National ways to the IWRM, or processes which hampered its implementation have emerged? In order to answer these questions the focus was on the institutional/organizational and operational structure of the state water authorities at the basin level, the district water departments, and the water users associations (WUAs) at the local level. Hence, three districts for each valley were chosen: Urgut, Nurabad, and Pastdargom (Samarkand province, Uzbekistan) and Tyulkibas, Ordabasy, and Otrar (South-Kazakhstan province, Kazakhstan). Focusing on the methodology, a comparative qualitative approach was used in order to collect the data: semi-structured interviews and informal talks were conducted to all the stakeholders (state bureaucrats, experts, WUAs' members) involved in the water management processes.

The evidence from the two case studies shows that the IWRM has not been implemented as initially sponsored by the international donors; Although with differences between the two states, the IWRM implementation was strongly influenced and shaped by local governments, somehow upsetting the framework's aims. Only the pillars which did not question and change the current water bureaucracies and related structures were selected and implemented, in order to achieve their national political-economic strategies; therefore two different national ways to the IWRM emerged. Based on a political geography perspective, this research is essential in in-depth understanding and in enhancing the debate on the IWRM implementation and related sociopolitical changes, in a region still influenced by the transitional processes which followed the collapse of the Soviet Union.

# SOMMARIO

Sin dall' antichità le risorse idriche hanno sempre svolto un ruolo strategico nello sviluppo dei territori e delle società, in particolare nelle regioni aride e semiaride; in queste aree del mondo infrastrutture idrauliche e sistemi di canalizzazione sono stati creati per estendere le aree irrigue e permettere lo sviluppo delle società, come è avvenuto lungo il Nilo, l'Indo e in Mesopotamia. Come evidenziato da Wittfogel (1957), in passato il rapporto tra gestione delle acque e l'emergere di sistemi politici dispotici e fortemente centralizzati hanno portato alla creazione delle *Società Idrauliche,* basate su un forte controllo sociale da parte dello Stato attraverso la gestione delle infrastrutture idrauliche.

Se durante il Novecento la gestione delle risorse idriche era considerata una problematica tecnica e ingegneristica sotto il controllo dello Stato e dei suoi tecnici (*Missione Idraulica,* Allan, 2001), negli ultimi decenni si sono verificati importanti cambiamenti: a causa di diverse problematiche socio-politiche e ambientali, i processi di gestione dell'acqua, in alcune parti del mondo, si sono orientati verso la decentralizzazione e la partecipazione, coinvolgendo vari attori sociali e le loro diverse razionalità. Tuttavia in altre parti, in particolare in alcuni paesi in via di sviluppo, questi processi sono stati caratterizzati da dinamiche politico-sociali più complesse, che hanno frenato le riforme.

Per far fronte a queste problematiche a partire dagli anni Novanta diverse agenzie e organizzazioni internazionali, tra cui le Nazioni Unite, la Banca Mondiale, la Banca Asiatica per lo Sviluppo, hanno iniziato a promuovere a livello globale, in particolare nei paesi in via di sviluppo, un nuovo paradigma nella gestione delle acque orientato al concetto di sostenibilità sotto una prospettiva ambientale, economica e socio-politica: l' Integrated Water Resources Management (IWRM). Basato sul dibattito tra tecnici e membri di organizzazioni internazionali che ha avuto luogo durante la Conferenza sull' Acqua e l'Ambiente di Dublino (1992), l'IWRM è strutturato attorno a diversi punti fondamentali al fine di migliorare, secondo un ottica sostenibile, la gestione delle acque. Attraverso l'IWRM le agenzie internazionali hanno inoltre cercato di promuovere i modelli dell'Irrigation Management Transfer (IMT) e la creazione delle Water Users Associations (WUAs) al fine di supportare la decentralizzazione e la partecipazione nei processi decisionali. Tuttavia nel corso dell'ultimo decennio è emerso un dibattito, prevalentemente composto da accademici, che si è interrogato sulla possibilità di implementare in modo efficace l'IWRM a livello globale, senza prendere in considerazione le diverse necessità e le diversità fisico-ambientali-politico e sociali dei vari Paesi.

L'obiettivo di questo studio è di comprendere ed analizzare i processi di implementazione dell'IWRM e le connesse problematiche di gestione dell' acqua nell' Asia Centrale post-sovietica, comparando due casi di studio: la Media Valle dello Zeravshan (Uzbekistan) e la valle dell'Arys (Kazakhstan). Dopo il crollo dell'URSS le nuove repubbliche indipendenti hanno dovuto colmare il vuoto politico- istituzionale e intraprendere un difficile processo di transizione e riforme della gestione dell'acqua. Risulta quindi necessario analizzare e comprendere quali logiche emergono dall'implementazione dell'IWRM (promosso nella regione dalla fine degli anni Novanta), se e come i due Paesi hanno influenzato il processo a seconda delle rispettive strategie ed obiettivi politico-economici e quali differenze o similarità sono emerse tra Uzbekistan e Kazakhstan. Per analizzare queste problematiche, l'attenzione si è focalizzata a livello locale su tre province (*rayon*) nella valle dello Zeravshan (Urgut, Nurabad e Pastdargom) e tre nella valle dell'Arys (Tyulkibas, Ordabasy e Otrar): prendendo i considerazione i punti fondamentali dell' IWRM, sono stati analizzati i meccanismi d' implementazione nelle agenzie di gestione dell'acqua a livello di bacino e nelle Water Users Associations (WUAs) a livello locale. Sotto il profilo metodologico i dati sono stati raccolti seguendo un approccio qualitativo tramite interviste semi-strutturate agli stakeholders coinvolti nei processi di gestione dell'acqua.

I risultati hanno evidenziato che l'IWRM non è stato implementato come promosso inizialmente dai donors: sebbene con differenze tra Uzbekistan e Kazakhstan, i governi dei due Paesi sono riusciti a influenzare e plasmare i processi d'implementazione dell'IWRM, orientandoli verso i proprio obiettivi politico economici nazionali in qualche modo contrastando i principi su cui il paradigma è costituito. E' possibile dunque affermare che due vie nazionali all'IWRM, fortemente influenzate dalle esistenti strutture politico-istituzionali, si sono delineate nella regione.

# INTRODUCTION – A Reflection on water and its policies

Since ancient times water resources have always played a strategic role in societal and territorial development. These dynamics have been even more strategic in the arid and semi-arid regions throughout the world due to variable water availability and potential scarcity issues; in these regions, rivers were diverted and canal systems were designed in order to increase the irrigated lands and to allow the development of societies, as occurred for instance along the Nile, Indo, Tigri and Eufrate valleys. In the past the relationship between water control and the emergence of strong despotic states led to the rise of the so-called *hydraulic societies,* based on the experiences arising within agricultural societies characterized by state-centralized water works.[1] In contrast to other natural resources, as mentioned also by Biswas (2008) and Sehring (2007), water is constantly in motion, flowing from one state to another, or from a particular natural region to another, making ownership claims challenging and sometimes involving conflicting processes; throughout the world, its availability, quantity, and quality significantly differ, leading to different and variable social or environmental issues.[2] Moreover water resources have been used and are currently used for different purposes, ranging from economic and technical to social and cultural concerns.

Therefore its management has always been a profoundly complex process, characterized and influenced by the different competencies and capacities of the involved institutions, sociopolitical conditions which affect the planning of water resources, and institutional and regulatory frameworks as well as different modes of governance. While during the twentieth century, water resources management was considered a technical issue mostly under the control of state-centralized bureaucracies and their hydro technicians and engineers, in the last decades a significant change has occurred. Due to different issues ranging from

[1] WITTFOGEL, K., 1957. *Oriental Despotism: a Comparative Study of Total Power*, Yale University Press, New Haven.

[2] BISWAS, A.K. 2008. Integrated Water Resources Management: Is it working?,*Water Resources Development,* vol. 24, n.1.

SEHRING, J., 2007. *The politics of Water Institutional Reform in Neo-Patrimonial States: a comparative analysis of Kyrgyzstan and Tajikistan,* PhD thesis, Fern Universitet in Hagen.

population growth and relevant political changes to land degradation and inequitable water access, it emerged that the management of water resources is not merely a technical issue, but asociopolitical, economic, and environmental one which involves a wide spectrum of different actors throughout the society and related strategies. Hence water management policies should be the result of debated strategies among the different stakeholders. Nevertheless, in several countries throughout the world, this change in rationale represented a significant and challenging issue, due to the different political and social contexts.

Based, as a starting point, on these water management challenges, this monography aims to analyze and understand these dynamics and specifically the Integrated Water Resources Management (IWRM) implementation processes, and its logic, in post-Soviet Central Asia — a wide heterogeneous arid and semi-arid region mostly included in the Aral Sea basin. These processes will be analysed in Uzbekistan and Kazakhstan, focusing at the local level according to a comparative approach. This research is essential in understanding institutional water reforms and related sociopolitical changes, according to a political geography perspective, in a region still influenced by the transitional processes which followed the collapse of the Soviet Union. In the next paragraphs the conceptual framework, the specific aims and research questions will be explained in depth.

Considering environmental and social issues in relation to water management and control, starting from the 1990s several international donors and development agencies such as the World Bank, the UN, USAID, the Asian Development Bank have sought to promote a new water management framework, both worldwide and in developing countriesin particular. Immediately, an initial question emerges: could a worldwide water framework be effectively promoted? Since the conference on "Water and the Environment", held in Dublin in 1992, where environmental and social issues in relation to water resources were discussed in terms of a sustainable perspective, the IWRM framework was launched.

Based on the so-called "Dublin principles" (see Chapter 1), the IWRM framework aims to improve water resources management according to multiple-perspectives sustainability: environmental, economical, social, and political ones. In order to implement the framework and achieve these aims, a guideline, characterized by different pillars, was designed; the guideline promotes the management of water resources according to territorial hydrographic boundaries (instead of administrative ones), the integration of the different water uses (irrigation, domestic use, and industry), the shift from a top-down vertical approach to a participatory horizontal one in the decision-making processes and the introduction of economic principles in water allocation services (water fees). Somehow, depending on the states, the pillars' implementation, would require institutional and organizational changes in the current national sociopolitical structures oriented to an adaptation to the IWRM. Since the end of the 1990s the Global Water Partnership (GWP) was created as the international agency oriented towards worldwide implementation of the IWRM as the new global water paradigm. Subsequently most of the international agencies seek to mainstream the framework through the establishment of different projects, in particular in developing countries, stressing the importance of reaching sustainable, efficient, equitable, and democratic use of water resources. The sponsor to the IWRM was integrated with the support of the following related inititatives: the Irrigation Management Transfer (IMT) and the establishment of the Water Users Associations (WUAs; see Chapter 1).

According to Allan (2003) this paradigm, which is defined as the political/institutional one, is the third sub-paradigm of the *reflexive modernity* phase in water management, initiated at the end of the 1970s with the environmental concern followed by the economic one at the end of the 1980s. This is considered a political paradigm because, according to his analysis, the IWRM framework implementation, requires institutional and structural reforms, and involves a political process to resolve potentially conflicting interests.[3] Nevertheless, as debated by Molle (2007), the agencies supporting the IWRM have tried to hide its evident political nature; in fact, as stressed also by Ghazouani et al. (2012) and Mollinga (2008), behind its support of multiple sustainability, the IWRM framework aims to roll-back state control of water resources and seeks widespread decentralization, liberalization, and the rise of private actors, as well as the introduction of economic and democratic principles withinpolitical-economic structures[4].

---

[3] ALLAN, T., 2001.IWRM/IWRAM: A new sanctioned discourse?, Occasional paper 50, SOAS/King's college University, London.

[4] MOLLE, F., 2008.Nirvana Concept, Narratives and Policy Models: Insights from the Water Sector, *Water Alternatives*, 1 (1).

GHAZOUANI, W. et al, 2012.*Water Users Association in the NEN region_IFAD interventions and overall dynamics*, IWMI, IFAD.

MOLLINGA, P., 2008. Water, Politics and Development: Framing a Political Sociology of Water Resources Management, *Water Alternatives*, 1(1).

In the last decade a wide debate on the IWRM among water professionals, donor members, and academia — in particular geographers, anthropologists, political scientists and hydraulic engineers— has emerged, focusing on and discussing its definition (GWP, 2000; see Chapter 1), its pillars, and in particular the implementation procedures and the benefits which it could lead to throughout the world. Regarding these last points specifically, various authors discussed whether the IWRM could be effectively implemented in all the designated countries, despite their different sociopolitical and economic structures, as well as their varying cultural backgrounds and environmental/physical features.

Reflecting on these issues, the following questions arise: how can the IWRM be put into practice in those states characterized by weak democratic structures or by authoritarian or semi-authoritarian political systems? It should be underlined that a considerable part of the developing countries are characterized by weak democratic institutions or by state-centric systems. Therefore, can the IWRM be effectively implemented in such contexts? Would the IWRM be shaped or influenced by the nation states' sociopolitical systems? In order to answer to these questions, it is necessary to take and in-depth look and analyze the current and former sociopolitical structures, meaning the strategies in conducting water reforms, the relations among the different actors involved in water processes, and the institutional and organizational structures of water organizations in a multiple perspective, considering the basin and the local level. In addition, the attitude of the water users is crucial in understanding these dynamics. As the concept of the IWRM is relatively new, there is a need for in-depth empirical research in order to understand the complexity of water resources management, the implementation of institutional reforms, the social relations between the state and the water users within its jurisdiction, and the effects of water policies on the territories. Data analysis from empirical case studies would allow a deep reflection and a potential generalization of IWRM's implementation processes and of the logic which influence these procedures as well as IMT performance. Therefore this research aims to make a contribution to the debate about the IWRM implementation processes in national contexts, enriching the field of water studies in the main framework of political geography.

Concerning the area, this research focuses on post-Soviet Central Asia, an arid and semi-arid region where water has always played a strategic role in societal development since ancient times; by diverting the flows of the two main rivers, Amu-Darja and Syr-Darja — flowing from the Tian-Shan and Pamir mountains to the Aral Sea — extended irrigated areas were designed in the last decades.

This region was chosen for the following reasons: firstly, since the collapse of the Soviet Union the newly independent countries, filling the institutional void, had to reform their agricultural and water sectors as well as their political and economic systems, which were inherited from the Soviet Union, concerning both sociopolitical and environmental issues. Secondly, due to this context, since the end of the 1990s, several donors and implementing agencies (the World Bank, Asian Development Bank, USAID and others) induced the independent states to apply and implement the IWRM framework and the related IMT in order to direct the water reforms towards a multiple sustainable perspective; in fact, the establishment of the WUAs were strongly promoted. Therefore, the research aim is also to contribute to Central Asian studies and to enrich the knowledge of the complex water resources management context which characterize this region. With the aim of a potential generalization of these processes in Central Asia, a comparative approach among two countries was chosen in order to highlight the similarities and differences in the IWRM's implementation procedures and to be able to answer the following questions: What are the logics which have affected the IWRM implementation? Were the national sociopolitical systems able to shape this process according to their strategies and aims? National policies to the IWRM, or processes which hampered its implementation have emerged?

Uzbekistan and Kazakhstan were chosen since they are the downstream countries where irrigated agriculture is more developed and plays an important role (30% GWP of Uzbekistan), and are the states with the largest water consumption of the whole Aral Sea basin. Since, as also stressed by Mollinga (2008), the local level is the scale where the implementation processes of national policies are more evident and understandable, this level was chosen for empirical research.[5] Therefore, in Uzbekistan the Middle Zeravshan valley was chosen as it is one of the most important irrigated areas of the country; while in Kazakhstan, the Arys valley, which is located in the southern part of the country, irrigated agriculture is mostly widespread. In order to answer the research questions, the IWRM pillars were taken into consideration (hydrographic management, integration and water users participation, and economic principles) before focusing on their current implementation level in the water authorities and organizations at the basin-local level. Therefore, the focus was on the institutional/organizational and operational structure of the state water authorities at

---

[5] MOLLINGA, P., 2008. "cit.".

the basin level, the district water departments, and the water users associations (WUAs) at the local level. Hence, three districts for each valley were chosen according to their physical location and territorial characteristics along the rivers from upstream to downstream: Urgut, Nurabad, and Pastdargom (Samarkand province, Uzbekistan) and Tyulkibas, Ordabasy, and Otrar (South-Kazakhstan province, Kazakhstan), (FIG.1).

*FIG.1: GIS elaboration a of a satellite image (NASA-Modis, 1999) which shows the two case-studies areas (yellow shapes) and the national borders (black lines).*

Focusing on the methodology, a comparative qualitative approach was used in order to collect the data. Since the research was mostly on social processes and connected dynamics this method was chosen because it allows a complete and deep understanding of them. Semi-structured interviews of the international and national expertswere conducted to understand specifically the institutional and organizational framework of water resources, followed by interviews with a wide range of stakeholders involved in water management processes at the basin and local levels: members and staff of the basin agencies, district water departments, and of the WUAs. Furthermore, both interviews and informal talks with the water users, peasant farmers, and household plot owners were conducted; in addition, field surveys were undertaken in order to understand the physical characteristics of the canal networks and related irrigated areas, and the water allocation procedures. Finally, some maps were designed, in order to show the case-studies areas, using and modifying satellite images and topographical maps through the ArcGIS 9 application.

*Outline of the book*

Following this Introduction, Chapter 1 presents an overview of the water management discourse and related paradigms in the last century, focusing in particular on the crisis of the *hydraulic mission* and the rise of *reflexive modernity* which led to the design of the IWRM framework. Then this concept, its pillars and related initiatives, IMT and related WUAs experience will be discussed in depth. This section is followed by a discussion of the scientific and academic debate about the IWRM rationale and its implementation processes. Chapter 2 focuses on the methodological approach, starting with an overview of the comparative qualitative methods in the social and water studies, followed by a description of the selected approach, the field research structure and its different phases. Chapter 3 presents the Central Asian water context starting with a regional geographical description highlighting the physical characteristics, the territorial changes in relation to water control, and the design of the irrigated areas. Afterwards, the water legacy and related strategies during the Soviet Union are discussed, followed by the regional institutional water and land reforms which affected the newly independent states during and after the Soviet collapse. Finally, the influence of the donors rationale and the initial widespread emergence of the IWRM and IMT in the Central Asian region is presented. Chapter 4 starts by presenting the water management structures and related reforms at the national level, comparing Uzbekistan and Kazakhstan, and follows with an in-depth analysis of the IWRM-sponsored water reforms, their strategies and rationales, at the basin level, in both countries. The subsequent chapters (5 and 6) focus on basin and local level case studies of Uzbekistan and Kazakhstan; the territories, the water authorities and organizations, and the IWRM-based institutional reforms are analysed and discussed, describing the water context and related issues in each of the selected districts. Chapter 7 compares the evidence and the results that emerge from the two case studies. The chapter is divided into different paragraphs: the first one compares the different interpretation of the water reforms and related paths to the IWRM at the basin level. The second paragraph focuses on the comparison of these processes at the local level for the twocase studies, discussing the implementation of each pillar of the IWRM framework. Then the fourth and the fifth paragraphs summarize and give conclusive remarks regarding the logics, the rationales, and the national interpretations of the IWRM. Finally, the last paragraph initially presents the lessons that can be learnt from the results of the research and from the IWRM implementation processes in the Central Asian region; then, coming back to the initial discussion, the findings are put in relation to the international debate presented at the beginning about the IWRM implementation and its procedures.

# 1. THE RULE OF WATER POLICIES: THE GLOBAL PARADIGMS FROM THE HYDRAULIC MISSION TO THE IWRM FRAMEWORK

## 1.1 THE HYDRAULIC MISSION AND THE DESIGN OF THE HYDRAULIC TERRITORIES

### 1.1.1 Introducing the Humans-Nature (water) relations

The importance of water resources management in dry lands cannot be overestimated, be it from a social, political, environmental, or economical perspective. Focusing on arid and semi-arid regions, from ancient times, human-environment relations — in particular river flows and agriculture-irrigation systems — have played a strategic role in the evolution of societies and territories. Reflecting on the Mesopotamia, the Nile or the Indus valleys, the relevancy of the state and societies' powers and strategies in controlling and managing water resources, with the aim of changing territories and creating irrigated areas,clearly emerges. Therefore the relations between (and among) the state and other actors and the environment are crucial points in order to understand these new territories, strictly connected with water, and their issues. As stated by Allan, Faggi, and other scholars, it is possible to define those new entities, which are the results of the interaction between society and nature, as *hydraulic territories,* since they are products of territorial transformations or reorganization through water carried out by the state.[6] According to Molle et al. (2009) these empires are famous for their success in controlling river systems and developing large-scale irrigated areas and agricultural production which supported and sustained their might and glory.[7] This is the case, for instance, in the development of the Nile delta oasis undertaken by the Egyptian empire, or the Mesopotamia irrigated area between the Tigri and Eufrate Rivers; in the Central Asian region too the development of the main oasis and related irrigated lands were carried out by strong political

---

[6] FAGGI, P.P., 1986. Pour un géographie des grands projets d'irrigation dans les terres seches des pays sous-developpè: les impacts sur le milieu et leurs consequences, *Revue de Geographié du Lyon,* vol. 61, n.1.

[7] MOLLE, F., MOLLINGA, P., WESTER, P., 2009. Hydraulic Bureaucracies and the Hydraulic Mission: Flows of Water, Flows of Power, *Water Alternatives,* 2(3).

powers, such as the Sogdian empire in ancient times and the Timur empire in the Middle Ages.[8]

### 1.1.2 The state, the people and the waters: the hydraulic societies

During the past centuries, water resources management and allocation policies throughout the world have changed and evolved reflecting particular interests and concerns, related both to national or international issues. In particular, in the arid and semi-arid regions, starting from the mid-1800s, water resource development carried out by the state has been an emergent and intentional political strategy to control space, water, and people, due to the increase of a positivist approach, engineering knowledge, and the ideology of the domination of nature.[9] The relation between water control and the emergence of strong despotic states formed the core of Wittfogel's research and analysis of the so-called *hydraulic societies* based on the experiences arising from the agricultural societies characterized by state-centralized water works. Wittfogel (1957) argues that the necessity to involve a big labour force to control water flows and establish irrigation networks and infrastructures was conducive to the development of a centralized bureaucratic type of state that he named "Oriental Despotism".[10] Hence, the water and labour force control led to the rise of a powerful elite characterized by scientists, engineers, and bureaucrats which achieved the technical and organizational knowledge to manage water resources. During the last decades the relations between irrigation facilities' control, state formation and centralized power have been strongly debated; nevertheless, the development based on water resources by state water bureaucracies has played a strategic role in state formation and power centralization, in particular in those regions where water is essential to conduct agriculture. As claimed by Molle et al. (2009), large-scale irrigation schemes construction and hydraulic works reappeared, — recalling those of the ancient empires — in the 1800s, in the main wave of colonialism, when colonial powers had the possibility to mobilise mass labour forces with the support of foreign engineers and related technical and organizational knowledge.[11] This trend could be observed in India and Egypt under the British Em-

---

[8] TOLSTOV, S.P., 1948. *Drevnii Khorezm*, Moskva, Izdanie MGU.

[9] ALLAN, T., 2003. IWRM/IWRAM: A new sanctioned discourse?, Occasional paper 50, SOAS/King's college University, London.

[10] WITTFOGEL, K., 1957. *Oriental Despotism: a Comparative Study of Total Power*, Yale University Press, New Haven.

[11] MOLLE, F., MOLLINGA, P., WESTER, P., 2009. "cit.".

pire or in the Central Asian part of the Russian Empire after 1870. It was properly at the beginning of the 1900s that large state investments became widespread, leading to the creation of water bureaucracies, which were supported by the positivist ideas of development through hydraulic engineering and irrigation, the domination of nature and the "let the desert bloom" utopia. A clear example of this modernist idea of controlling and changing nature through water management emerges on examination of the Hungry Steppe region, today included in Uzbekistan, which was transformed from a steppe area into an irrigated plain by the Tsarist engineers (at the beginning of the 1900s) and later (1960s) by the Soviet ones.[12] Nevertheless both the ideas/rationales and practices of water management partly differed according to the different regions and among the so-called "North" –meaning the Western countries and the Soviet empire- and "South"- meaning Africa, South-America and South-Asia-; Allan (2001) and other scholars use these terms to indicate and distinguish the developed from the developing countries, although it has been widely debated how it could be possible to separate all the countries of the world into two blocks, considering the heterogeneity among them. For these reasons often the term "plural" is added.

### 1.1.3 Approaching the water paradigms: the Pre-modern communities

According to Allan (2003), in the last two centuries several shifts in water paradigms throughout the world have occurred, due to several technical, political, and social issues; these changes of policies were more frequent in the North in comparison with the South.[13] The term "water paradigm" includes the rationale and the way of managing water resources which is influenced by the sociopolitical and economical discourse; in turn, the discourse is influenced and affected by the knowledge, narratives, and concepts supported by the scientific communities, governments, or international organizations. Awareness of scarcity and declining water quality have tended to increase the prominence and intensity of water policy–making and related debate. Starting from the 1800s Allan identifies five different water paradigms which globally affected water resources management; according to his analysis and classification, the first paradigm is associated with Premodern communities.[14] Explaining this paradigm he mentions the communities affected by limited technical and organizational capacities in natu-

---

[12] BICHSEL, C., 2012. "The Drought Does Not Cause Fear": Irrigation History in Central Asia through James C. Scott's lenses, *Revue d'Etudes Comparatives Est-Ouest,* vol.43, n.1-2.

[13] ALLAN, T., 2003. "cit.".

[14] ALLAN, T., 2003. "cit.".

ral resources management and in particular concerning the water sector. Nevertheless, this first paradigm is not explained in details; what emerges in Allan's reflection is that the way of managing water resources and developing water policies precedes the ideas of science and nature control —that's why it is named pre-modern. By the way, in analysing the history of water management and the relations between the pre-modern states/societies and arid territories throughout the world, as highlighted above, it emerged that in ancient times there already were relevant examples of organized societies which reorganized territories through the construction of water infrastructures and irrigation canals. The examples of the Nile River, the Mesopotamia, the Nabatean empires in present Jordan, the Chinese empire along the Yellow and Yangtze rivers, show that, although the ideas of positivism and modernity were a long way from being developed, these societies were already able to organize new *hydraulic territories* due to their organizational levels. In addition, what Allan described as pre-modern communities with limited technical and organizational capacities, partly clashes with what is stated by Molle et al. (2009) regarding empires which possessed great organizational capacities to manage water resources somehow reflect their might and glory.[15]

### 1.1.4 The paradigm of modernity: the hydraulic mission

Most scholars seem to agree that the end of the 1800s coincides with the rise of industrial modernity and its *hydraulic mission*. According to Allan's classification, the second paradigm is associated with industrial modernity and later with the *hydraulic mission* and the construction of the *hydraulic territories*. As mentioned above, industrial modernity in the water sector was featured by the Enlightenment, the sciences, engineering capacities, the belief that nature can be controlled, and the investments of the state and the private sector in water management. This paradigm affected the water sector development of both the Western capitalist economies (the United States of America, first of all) and later the socialist-planned economies guided by the Soviet Union. In fact, this phase was possible because of the revolutions in science and industry at the beginning of the 1900s, the achievements of capitalist organization in dealing with labour, environment, and capital resources, and later by the socialist countries in dealing with planned state economies and socialist labour. It is hence possible to state that this new paradigm was formulated in the "North" (that is, in Western capi-

---

[15] MOLLE F., MOLLINGA, P., WESTER, P., 2009. "cit.".

talist countries and then in socialist ones) and later (starting from the 1950s) was transferred to the plural "South" through a knowledge flow which affected some countries and their colonies (for instance, France and northern Africa) or other countries and their former colonies or allies (such as the United Kingdom with Pakistan and India, and China with Vietnam). Hence, since the 1950s–1960s this water resources management approach and related policies started to be implemented in a plural way by the independent governments of the republics of the "South". Therefore this approach in managing water resources and reorganizing territories in terms of a productivity perspective required a structured and organized state bureaucracy. According to Molle et al. (2009), supported with the legitimacy of new techniques (such as dams and hydropower) and the unlimited power of science, inspired by the mission to make the deserts and the steppes bloom, hydraulic bureaucracies were created with the aim of facing the challenges of flood protection and large-scale public irrigation; these bureaucracies acted in the name of the common good, in relation with politicians and national leaders.[16] The *hydraulic mission* was born when the idea of "not a single drop of water should reach the sea without leading benefits to the communities and societies" emerged. Hence, the idea of the *hydraulic mission* was to change water use, oriented towards a maximization of the profits in agriculture, to the increase of irrigated areas leading to benefits to the communities through the construction of dams and irrigation schemes, and generally to demonstrate that the role of science and new techniques in controlling nature could lead to relevant benefits to the state economy and to the social system; in addition, as mentioned by several scholars, this approach to water management leads to a strong legitimization of the state and its hydraulic bureaucracies towards the population in managing natural resources (FIG.2).[17]

---

[16] MOLLE F., MOLLINGA, P., WESTER, P., 2009. "cit.".

[17] FAGGI, P.P, 1986. "cit.".
    ALLAN, T., 2003. "cit".
    RAFFESTIN, C., 1981. *Pour une Geographie du Pouvoir,* Paris, ed. LI TEC.
    AMINOVA, M., ABDULLAEV, I., 2009. "cit.".

14

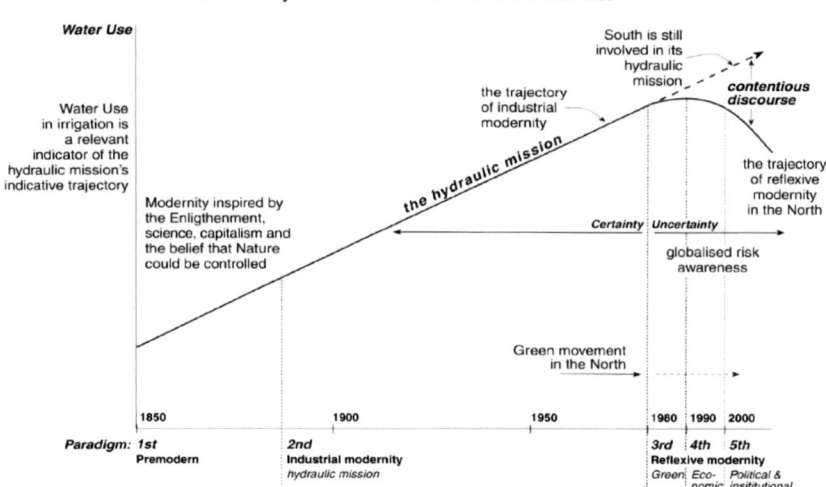

Neo-liberal modernity and the water sector in semi-arid countries

*FIG. 2: Scheme "Neo-liberal modernity and the water sector in semi-arid countries" (source: Allan, T., 2003. "cit.".) which shows the water sector and the water paradigms in the last 150 years in semi-arid countries and the differences and the gaps between the so-called "North" and "South".*

Relating the hydraulic mission to territorial reorganizations, the research of Faggi (2002), influenced by the Turco (1988) and Raffestin (1981) on the territory and territorialisation processes, highlighted the concept of the *"hydraulic territorialization"*. According to Faggi, *hydraulic territorialization* is a process of construction, production, and management of a certain territory, affected by water scarcity, through the use of the irrigation water and the mediation of the actors involved in the process.[18] In relation with the hydraulic mission phase, Faggi identifies the state policy characterized by strong procedures in water management, a strategic hydraulic policy which reflects its legitimization, and the construction of huge water facilities.[19] This approach clearly reflects the steps and aims of the hydraulic mission and the procedures of the hydraulic bureaucracies.

---

[18] RAFFESTIN, C., 1981. "cit.".

  TURCO, A., 1988. *Verso una teoria geografica della complessità,* Unicopli, Milano.

[19] FAGGI, P.P., 2002. La valle del Sourou (Burkina Faso): per una geografia della territorializzazione in Africa, *Rivista Geografica Italiana,* vol. 109, n.2.

## *1.1.5 The heterogeneous trajectories of the hydraulic mission*

At the beginning of the 1900s political processes led to the creation of hydraulic bureaucracies in different parts of the world; for instance, the US Bureau of Reclamation was created in the USA, the National Irrigation Commission in Mexico, and the General Directorate of Public Works in Turkey. In Europe part of these state offices for water management had been previously created, but in this period they gained more relevancy with the construction of new canals, hydropower plants, and dams. Starting from the 1930s in the Soviet Union, appeals for large-scale hydraulic projects had arisen, according to the main idea that technology and mechanization would be the solutions for economic and social issues with the state's main vision of the "supremely rational society". Under the slogan "we will instruct nature and we will receive freedom", Stalin's bulldozer technology planned massive hydraulic works including the damming of the Volga River and the White Sea-Baltic canals, using the forces of masses of slaves and workers from the recently established collective and state farms.[20] The emphasis on massive hydraulic works and water development projects increased in the 1950s–1960s due to the reconstruction after the end of the Second World War and to the independence of several former colonies in Africa and Asia. These processes were fuelled by the search for national symbols and national ways of development and natural resources management in progress during the Cold War and therefore featured, depending on the different countries, by the Soviet Union's influence or by the USA's. These processes and this path to new national identities were more relevant in the countries which had recently attained their independence. As pointed out by Molle et al. (2009), these processes led to three different but related forms of the hydraulic mission that combined to give way to its apogee: first, a re-interpretation of the ancient "Oriental Despotism", in terms of strong state legitimization and related policies in water resources management, in the Soviet Union and in the People's Republic of China; second, a state-led massive capital investment in hydropower dams in the Western countries and also in irrigation in countries affected by water scarcity, like Spain, Australia, and western USA; and third, a "post-colonial despotism", in terms of a national reinterpretation of the hydraulic mission, in parts of the newly independent countries from the "South".[21] In the Soviet Union the

---

[20] JOSEPHSON, P.R., 1995. "The Project of the Century" in the Soviet Union: Large-scale technologies from Lenin to Gorbachev, *Technologies and Culture,* 36:3.

[21] MOLLE F., MOLLINGA, P., WESTER, P., 2009. "cit.".

plans for the transformation of nature and the creation of new hydraulic territories to reinforce the idea of the *"homo sovieticus"* and that one of the socialist labour, reached in this period the apogee. In the arid Soviet Central Asia, the Fergana and the Karakum canals, which are the longest in the world, spanning 1375 kilometers, irrigating Turkmen SSR, were built in the 1940s by the Soviet workers with the ideological support of science and engineering and socialist labour; under Kruscev, starting from 1953, the Virgin Lands' Plan was launched, with the aim of reorganizing part of the steppes in north-eastern Kazakh SSR into fertile agricultural lands for wheat and corn farming.[22] In the 1960s–1970s another big project was under consideration, "the Siberian Rivers Reversal", with the aim of diverting the water flow of the Siberian rivers (Irtysh, Ishim, and others) into the irrigated areas of Central Asia; this project could be considered a real utopia in terms of the hydraulic mission paradigm, but it has never been implemented due to the financial costs and the natural risks. During the same decades, and influenced by a close ideological approach, several projects for hydropower and river flood control were launched and implemented under the advice of Soviet engineers in the People's Republic of China. Focusing on the second approach of the hydraulic mission (the western one pointed out above), it is possible to recognize similar approaches in the US during the 1950s, with the aim of demonstrating to the world the superiority of the capitalist system in the Cold War: in response to the Siberian Rivers Reversal, a gigantic project to divert water from Alaska to the arid west and then to Mexico was launched.[23] In Europe, before the Second World War, huge projects were launched in Spain and Italy during Franco and Mussolini's dictatorships, mostly for dams in Spain and wetlands reclamations in Italy, to legitimize their strong power towards the rural elites.[24] In a parallel way, the governments in the "South", in order to strengthen nation building and to legitimize their new powers, embraced the icons of modernity and development through land reclamation, territory reorganization, and irrigation schemes construction. In these processes they were supported both by the US and the Soviet Union, depending on the independent countries' position

---

[22] WEGERICH, K., 2008. Hydro-Hegemony in the Amu-Darja Basin, *Water Policy,* 26(2).

O'HARA, S., HANNAN, T., 1999. Irrigation and Water management in Turkmenistan: Past Systems, Present Problems and Future Scenarios, *Europe-Asia Studies,* vol. 51, n.1.

[23] JOSEPHSON, P.R., 2002. *Industrialized nature: Brute force technology and the transformation of natural world,* Washington DC, Island Press.

[24] PIASTRA, S., 2012. *Land Reclamation: Geo-Historical Issues in a Global Perspective,* Proceedings of the International Conference held at the University of Bologna, Patron ed.

in the Cold War, which had economical and political interests in influencing this post-colonial hydraulic mission.[25] In these countries — for instance India, part of the Middle East and the Sahel belt — still agriculturally based and hence with the aim of restructuring water facilities and achieving food security, the hydraulic mission was the best approach, according to the new bureaucracies, to undertake the development processes. Throughout the world the decades following the Second War World were therefore characterized by a strengthening of different forms of the hydraulic mission and an empowerment of the hydraulic bureaucracies which were able to strengthen their legitimacy. In post-war Vietnam, as claimed by Evers and Benedikter (2009), the state reused the wartime propaganda to induce the people to conduct massive works in a collective way to reorganize territories through hydraulic works and water management; this social work and cohesion contributed to reinforcing the idea of nation building.[26] In the 1960s and 1970s, the hydraulic mission — in particular in Western countries due to democratic systems and private companies' involvement — was not only taken on in the interest of governments but also welcomed in synergy with other actors like construction companies, development banks, and businessmen.[27] Therefore, relations of interests and flows of money between the government, local politicians, and banks' heads became widespread and influenced water policies and the trajectories of the hydraulic mission. It must be stated that throughout the world the projects and the pharaonic structures of the hydraulic mission have not only led to legitimization and benefits for bureaucracies and populations but also to several environmental problems.

---

[25] BETHEMONT, J., FAGGI, P.P., ZOUNGRANA T.P., 2003. *La Vallé du Sourou (Burkina Faso): Genese d'un territorie hydraulique dans l'Afrique Soudano-Sahelienne,* l'Harmattan France.

[26] EVERS, H.D., BENEDIKTER, S., 2009. Hydralic bureaucracy in a modern hydraulic society: Strategic group formation in the Mekong Delta, Vietnam, *Water Alternatives,* 2(3).

[27] MOLLE F., MOLLINGA, P., WESTER, P., 2009. "cit.".

## 1.2 THE SHIFT IN WATER PARADIGMS: THE RISE OF THE REFLEXIVE MODERNITY

### *1.2.1 The hydraulic mission in crisis in the "North"*

The idea of controlling and dominating nature through science and new technologies did not take into account the environmental issues caused by rivers' diversions and monoculture agriculture, such as soil degradation and salinization and the desiccation of river beds and lakes; the shrinking of the Aral Sea, which started in the 1960s, is an evident effect of these water policies. Therefore, in the mid-1970s, the paradigm of the hydraulic mission started to be debated and challenged: the notion that water resources were being damaged rather than controlled by the impact of the alliance of science, technologies, and national investments gained currency in the "North" and in the "northern" development banks and agencies. Furthermore, a reflection on the negative pressure on the environment and on the damages caused rather than the real benefits of industrial modernity had already been considered by environmentalists such as Carson (1965) during the apogee of the hydraulic mission.[28] Therefore this debate had led to increasing calls for an "ecologic turn" in water management, with more emphasis placed on "putting water back into the environment".[29] But these debates and reflections have not involved all the countries of world, but mostly the nations of the "North"; hence, as it is possible to see in the previous scheme, during the mid-1970s a gap or a contentious discourse, as identified by Allan (2001), emerged between the "North" and the "South". As this author mentioned, in the USA the environmental principles in water and natural resources management entered the agenda through the actions of President Jimmy Carter, who challenged the political networks and the procedures put in place in the previous decades with the aim of strongly promoting a shift towards a more environmental-friendly approach in natural resources management.[30] Nevertheless, this strong support towards a change in the water paradigm and related policies mostly came from the activists movements, ecologists and scientists, mostly based in the USA and in Western Europe. In addition, in this period the first international conferences and meetings focusing on the environment—for instance, the one in Mar de Plata (Argentina) in 1977 — were organized to discuss the uncertain future of natural resources management.

---

[28] CARSON, R., 1965. *Silent Spring,* London, Penguin Books.
[29] MOLLE F., MOLLINGA, P., WESTER, P., 2009. "cit.".
[30] ALLAN, T., 2003. "cit.".

## *1.2.2 The environmental concerns and the support to the reflexive modernity*

According to Allan (2001), these processes and debates in the mid-1970s led to a shift from industrial modernity, which lasted almost a century, to a phase classified as reflexive modernity; in the "North" this phase has been featured by three different water management sub-paradigms[31]:

- Since the mid-1970s: the Green paradigm
- Since the 1990s: the Economic paradigm
- Since the 2000s: the Political paradigm

The first water sub-paradigm of reflexive modernity can be defined as the green paradigm since it represented and was characterized by a change of water use and management priorities inspired by the environmental awareness of the green movements. These grassroots movements, succeeded in influencing governments in shifting their water policies towards a more sustainable approach, reducing — particularly in industrialized semi-arid countries — water use for irrigated agriculture and increasing water allocation back to the natural "hydrocycle". Furthermore, at the end of the 1970s and beginning of the 1980s, in several European countries and the US, due to scientific debates and to the action of the green movements, ministries of the environment were created; hence, environmental scientists and representatives of these movements found a place in the political power structures. As stated by Molle et al. (2009), the hydraulic bureaucracies, which had governed the water sector until the 1970s, were challenged by civil society which opposed various big hydraulic works, like dam construction or river divertion; in addition, they were also challenged by internal divisions, due to different interests, ideas, and perspectives of future water management in that phase of uncertainty and change.[32] In fact, Allan clearly claims that this significant shift from industrial modernity to reflexive modernity coincided with a shift from certainty — in terms of well-known procedures in water processes — to a phase of uncertainty. Nevertheless, at the beginning of the 1980s this environmental approach led to relevant changes in water policies in parts of the "North" industrialized countries like Israel, Australia, and some states of the USA (such as, California and Arizona). This shift in water policies also affected the attitudes which had formerly characterized hydraulic works and

---

[31] ALLAN, T., 2003. "cit.".
[32] MOLLE F., MOLLINGA, P., WESTER, P., 2009. "cit.".

water facilities construction: according to the reflexive modernity the paradigm was oriented towards control and maintenance of existing infrastructures instead of design projects to create new ones. As mentioned before, a notable gap and contentious discourse emerged in this phase between the "North" and the "South". Although considered part of the "North", the Soviet Union in the 1970s and 1980s had not been affected by these reflections and changes in water policies and went on carrying out its hydraulic mission both in the European Russian plains, with the construction of new dams and canals, and in the Soviet Central Asian republics; for instance, in this period the irrigated area of the Fergana valley (Uzbek-Kyrgyz-Tajik SSRs) increased, as well as that of the Hungry Steppe (Uzbek SSR) reorganized in the Syr-Darja irrigated plain.[33] As previously stated, the "South" too, being obviously unaffected by the debate on environmental issues, continued carrying out its hydraulic mission featured by the main goals of development, poverty alleviation, and the extension of irrigation schemes.

### *1.2.3 Towards a commodification of water: the Economic sub-paradigm*

During the 1980s, and specifically at the end of this decade, the western "North" was affected by a strengthening of the 1970s' approach and by further changes in water resources management and related policies; these changes were mostly oriented towards an economical perspective. According to Allan (2003), this second sub-paradigm of reflexive modernity, was named the Economic sub-paradigm. The reason is that this further change in water resources management approach was inspired by economists and economic scientists. They had drawn the attention of governments and water users in the western "North" to the "economic value of the water" and its importance as a scarce economic good; these ideas were considered and gained currency in particular between the 1980s and the 1990s.[34] Since it was the first time that water resources were effectively considered a commodity- although it had already been questioned before the issue of water saving had become a relevant concern — this statement has fuelled a strong debate both at national level and at the international or global level. The most debated topics concerned the system of water pricing, the potential decentralization and privatization of water management and allocation, and the role

---

[33] BICHSEL, C., 2012. "cit.".
   BICHSEL, C., 2011. Liquid Challenges: contested water in Central Asia, *Sustainable Development Law and Policy* .
[34] ALLAN, T., 2003. "cit.".

of the international organizations and development banks in these processes. Furthermore, reflection on this economic sub-paradigm has focused also on the existing gap between the "North" and the "South" and how to spread these narratives (Molle, 2008) in countries still partly involved in the hydraulic mission.[35] It should be stated that these issues and related debates have emerged in a particular sociopolitical and economic phase, between the 1980s and 1990s, which means the end of the Cold War and the subsequent ascent of the capitalist system and its related political and economic features as the dominant model. Although the water pricing system was already widespread in Western countries, on the one hand, the idea of "water as an economic good" was quite well accepted because it would lead to more efficiency in water allocation oriented towards water saving, but on the other hand, it was strongly debated because of the risk to become, through a privatization process, a market-oriented commodity; in addition, the grassroots movements, which have supported the environmental reflection on natural resources management, were reluctant to adopt the idea of the privatization of services and of water resources as a market commodity. [36] In the "South", affected by the challenges of development, poverty alleviation, and food and drinking water security, the idea of water as an economic good and subsequent issues about water pricing were initially considered as a narrative elaborated in the "North", blind to the developing countries' issues.

### 1.2.4 Decentralizing water management

Focusing on the political systems, in relation to natural resources management, this phase, as mentioned before, was affected by a rethinking of the role of the state in water control, a decentralization oriented to the support and promotion of the local water-users communities' involvement, and the emergence of the private sector and development banks. According to Molle et al. (2008), the financial squeeze which affected several countries at the end of the 1980s, state budgets under financial pressure and structural adjustment programs, were pushed towards a strong reduction of the large inflow of money which had fuelled the water resources development in the previous decades.[37] Already two decades before, Hunt (1989) stressed that the large-scale public irrigation investments promoted until the 1980s had not achieved the increases in productiv-

---

[35] MOLLE, F., 2008. "cit.".

[36] RAHAMAN, M., VARIS, O., 2005.Integrated Water Resources Management: Evolution, Prospects and Future Challenges, *Sustainability, Science, Practice and Policies,* vol.1, n.1.

[37] MOLLE F., MOLLINGA P.,WESTER, P., 2009. "cit.".

ity which were expected, in particularly in the developing countries.[38] According to the same issue, Ghazouani et al. (2012) claimed that due to the reduction of several state budgets for water resources management, part of the hydraulic infrastructures, both at primary and local levels, started to deteriorate; hence, a reassessment of the water sector and related policies was necessary.[39] These issues mostly affected the Soviet Union and the socialist republics of Eastern Europe at the end of the 1980s and part of African and Asian countries. Hence, those issues, together with a neo-liberal criticism of state water management, led to a reform of the water sectorin several countries, characterized by a decrease of subsidies, schemes rehabilitation, decentralization processes, water price-fixing frameworks, and community development; these policies, as mentioned above, emerged in a neo-liberal context of structural adjustment and broke away from the idea that water resources have to be exclusively managed by the state and its institutions.[40] Molle et al. (2008) added that another relevant strategy included in this water reform framework was to shift part of the water costs from the state to the water users, and also to strengthen the idea of water saving, supported both by environmental and economic concerns. Focusing on the Sahelian region, Faggi argues that at the end of the 1980s in several countries the ongoing hydraulic mission fell into crisis and a structural adjustment of the water sector and related policies was necessary to improve irrigation schemes' operation & maintenance (O&M), transferring the management control from state departments to the local communities of water users.[41]

### 1.2.5 The rise of international agencies in water policy-making

This important wave of political and institutional changes was caught by different international agencies and banks which supported in various ways the roll-back of the state and were therefore able to strengthen their roles and actions; the most influent donors like the World Bank (WB), the Asian Development Bank (ADB), the United States Agency for International Development (USAID), the United Nations (UN) the Swiss Development and Cooperation

---

[38] HUNT, R., 1989. Appropriate Social Organization? Water Users Associations in Bureaucratic Canal Irrigation System, *Spring*, vol.48.

[39] GHAZOUANI, W., MOLLE, F., RAP. E., 2012. Water Users Associations in the NEEN region - IFAD interventions and overall dynamics.draft, submitted to IFAD.

[40] GHAZOUANI,W., MOLLE, F., RAP., E., 2012. "cit.".

[41] FAGGI, P.P. et al., 1995. *Irrigazione, Stato e Territorio in Sudan; il gioco della posta in gioco*, Terra d'Africa, Milano, Unicopli.

Agency (SDC) and others started to promote, in several developing countries, international projects focusing on different topics related to water and development in order to support and influence water practices and policies reforms. The collapse of the Soviet Union in 1991 and its subsequent political, institutional, and organizational emptiness gave international donors the possibility for widespread dissemination of their approach in water and natural resources management in the newly independent countries, which featured a roll-back of the state, decentralization, and involvement of the water users in the decision-making processes.[42] Generally, as Allan (2003) states, international donors assumed the responsibility to globally extend the "reflexive modernity" paradigm, particularly in the countries of the "South".[43] In addition, the independence of new countries and the fall of the socialist ideology in African and Asian countries outside the Soviet Union inevitably led critical issues be solved in the wide field of natural resources management, such as food and drinking water security, infrastructure deterioration, farmlands reclamation, and poverty alleviation. Therefore, in part of those countries, in particular in those affected by the hydraulic mission until the 1990s, water resources management and allocation shifted from a purely technical issue carried out by a centralized state to a sociopolitical and economic concern.[44]

### *1.2.6 The reform of water management as a multi-perspective challenge*

At the beginning of the 1990s it became increasingly evident that the water issues of a country could not be resolved only by the water professionals and ministers alone; as mentioned above, water problems started to become interconnected and interlinked with other development-related issues and also with environmental, economic and sociopolitical concerns at local and national levels and sometimes at the international level, involving the independence of new republics,[45] Moreover, as mentioned by the Global Water Partnership (GWP, 2009), the trend of the last two decades indicates that water problems in the near future will be more and more interconnected with energy supply, agriculture, and in-

---

[42] AMINOVA, L., ABDULLAEV, I., 2009. Water Management in State-Centered Environment: Water Governance analysis of Uzbekistan, *Sustainability*, 1.

[43] ALLAN, T., 2003. "cit.".

[44] ABDULLAEV, I., MOLLINGA, P., 2010. The Socio-Technical Aspects of Water Management: Emerging Trends at Grass Roots Level in Uzbekistan, *Water*, 2.

[45] BISWAS, A.K., 2008. Integrated Water Resource Management: Is it working?, *Water Resources Development*, vol. 24, n.1.

dustry and with social sectors such as education, environment, and regional and local development.[46] It has been therefore widely argued that the goal of water management for the future will be not only concerned with technical improvements, but something strictly related with regional and local development and the improvement of livelihoods systems. According to the debate among water professionals, scientists, and international development banks, this new perspective of water management required a shift from state-centred policies to society-centred ones.[47] Therefore, to achieve the above-mentioned goals through a wide and multi-disciplinary and multi-sectoral water management perspective, the term "integration"— in reference to integration of water issues with agricultural, industrial, and energy issues, for example — emerged. In addition, this idea of facing global development issues and food and water security through a rethinking of water management in a multi-disciplinary perspective was fuelled in 1992, in preparing for the Rio International Conference on Environment and Development, by the idea that environmental issues, ecosystems protection, and water security had become global concerns.[48]

## 1.3 THE IWRM FRAMEWORK AND ITS INITIATIVES

### 1.3.1 An integrated approach to globally address the water issues

Due to the analysed global sociopolitical and economic changes and to the unprecedented management complexities, several specialists in the water community (international agencies, development banks, academic departments) started to look for a new paradigm for water management which could solve the existing issues in different parts of the world.[49] As the GWP experts (2001) claimed, the world's water resources were and also are under increasing pressure due to population growth, increased economic activities, and generally to an improved standard of living which has led to an increased competition for the limited freshwater resources. From another perspective, social inequities, economic marginalization, and poverty has led to soil and forest over exploitation which

---

[46] GLOBAL WATER PARTNERSHIP, 2009. *A Handbook for Integrated Water Resources Management in Basins,* Elanders, Sweden.

[47] MOLLINGA, P., 2007. *Water Policy-Water Politics: Social Engineering and Strategic Action in Water sector Reforms,* ZEF Working Paper series, n.19. University of Bonn.

[48] ALLAN, T., 2003."cit.".

[49] BISWAS, A.K., 2008. "cit.".

has often affected  water resources in a negative way.[50] These issues needed to find appropriate ways to coordinate policy-making, planning, and implementation in an integrated manner across sectoral, institutional, and professional boundaries. Moreover, this new paradigm needed to shape future water policies throughout the world and balance the existing relevant differences among the various countries in order to promulgate the concept of sustainability. Therefore the experts of the water community stressed that a comprehensive and integrated water resources management was needed for the following reasons:[51]

- Limited fresh water resources are becoming more and more polluted, rendering them unfit for human consumption and also unfit to sustain the ecosystem
- Those limited water resources have to be divided amongst the competing needs and demands in society
- Many citizens do not as yet have access to sufficient and safe fresh water resources
- Techniques used to control water (dams and dikes) may often have undesirable consequences on the environment
- There is an intimate relationship between groundwater and surface water, between coastal water and fresh water, etc. Regulating one system and not the others, without the supported integration, may not achieved the desired results

In analysing these issues, it emerged that an integrated approach in water resources management, together with an institutional change towards sustainability would lead to wide benefits and would better address current global problems, such as water pollution, social equity in water access, and environmental degradation due to the misguided policies undertaken in previous decades. Hence, economical, social, ecological, and legal aspects need to be considered, as well as quantitative and qualitative aspects.

---

[50] GUMBO, B., VAN DER ZAAG, P., 2001. *Principles of Integrated Water Resources Management (IWRM),* Global Water Partnership (GWP) Southern Africa, Southern Africa Youth Forum, 24-25 September, Harare, Zimbabwe.
[51] GUMBO, B., VAN DER ZAAG, P., 2001. "cit.".

## *1.3.2 The Dublin principles and the design of the IWRM framework*

In January 1992, during the International Conference on Water and the Environment held in Dublin (Ireland) to prepare the water issues agenda for the UN Rio conference, a new framework for water management was formulated: that is, the Integrated Water Resources Management (IWRM), based on the following four guiding principles:[52]

- Principle one recognizes fresh water as a finite, vulnerable resource, essential to sustain life, development and the environment; it should be managed in an integrated manner

- Principle two recognizes that water development and management should be based on a participatory approach, involving water users, planners and policy-makers at all the levels; water should be managed with the people and close to the people

- Principle three recognizes that women play a central role in the provision, management, and safeguarding of water; involve women in all the processes

- Principle four recognizes that water has an economic value in all its competing uses and should be recognised as an economic good; ensure basic human needs through water allocation and move towards full cost pricing to encourage rational use and recover costs.

Beside the support to the integrate approach stressed by water community experts, the four guiding lines discussed in Dublin focused, for the first time on the "official" water debate, on the relevancy of the participatory approach in the decision-making processes, in order to involve all the stakeholders including at the farm level, and in particular highlighting the central role of women in water management. Furthermore, the previously debated concept of water as an eco-

---

[52] GWP, TAC, 1998. *IWRM- At a glance,* GWP documents.

nomic good, was recognized and formalized as a move to encourage a rational use against wastes and a cost recovery for independent water users organizations. The initial four Dublin principles were associated with the following list of key concepts: [53]

- Integrated water resources management, implying:

    1. An intersectoral approach
    2. Representation of all the stakeholders
    3. Consideration of all the physical aspects of water resources
    4. Considerations of sustainability and the environment

- Sustainable development, sound socio-economic development that safeguards the resource base for future generations
- Emphasis on demand driven and demand oriented approaches
- Decision-making at the lowest possible level (subsidiarity)

Decision-making would involve the integration of the different objectives where possible, and a trade-off between these objectives where necessary, according to societal aims. The design of the Dublin Conference principles had a strong impact in the water-professionals debate and also in governments and international donors' opinions. Incidentally, it had a stronger impact in the "North" compared to the "South" since it was held in Europe and was promoted mostly by "Western" actors. The main successes of the Dublin conference were that it focused on the necessity of integrated water management and on active participation of the stakeholders, from the highest levels of government to the smallest communities, and highlighted the special role of women in water management; those recommendations were later consolidated into Chapter Eighteen of Agenda 21 in the 1992 Rio de Janeiro conference.[54] Nevertheless, as claimed by Rahaman and Varis (2005), the Dublin conference was criticized because it mostly included water experts without paying attention to the inclusion of other governmental and non-governmental stakeholders; moreover, some governments and water professionals of the developing countries not only criticized the Dublin principles, in particular the economic dimension of water, based on the second sub-paradigm of the reflexive modernity (Allan, 2003), but also the fact that no ade-

---

[53] GUMBO, B., VAN DER ZAAG, P., 2001. "cit.".
[54] RAHAMAN, M., VARIS, O., 2005. "cit."

quate guidelines were provided to implement them in the complex water scenario in developing countries.[55] A rethinking of the relations between national and international actors and of integration in natural resources management was necessary.

### 1.3.3 The IWRM into practice: the role of the Global Water Partnership

This new approach required a holistic perspective and subsequently an unprecedented level of political and international cooperation in order to strengthen the integrated approach. Therefore several governments, international donors and agencies started to question how to manage those principles and how to put these statements into practice, creating a process of water policies reform oriented to Integrated Water Resources Management.[56] With the aim of providing a guideline for its implementation, in 1996 the Global Water Partnership (GWP) was created to foster the development of a road map for Integrated Water Resources Management. GWP was designed as an international network open to all the organizations dealing with water resources management, hence, including government members of developed and developing countries, agencies of the United Nations, development banks and donors as well as academic and water professionals' networks, and the private sector[57]. Furthermore, the GWP governance includes the Technical Advisory Committee (TAC), featuring scientists and professionals, charged with developing an analytical framework of the water sector and proposing actions which will promote sustainable water resources management. Therefore, a framework based on the Dublin principles and incorporating all four dimensions of efficient water management (social, environmental, economic, and political) was provided; below are the guidelines: [58]

---

[55] RAHAMAN, M., VARIS, O., 2005. "cit."

[56] SNELLEN, W.B., A. SCHREVEL, 2004. *IWRM: for sustainable use of water: 50 years of international experience with the concept of Integrated Water Management,* Background document to the FAO, Netherlands Conference on Water for Food and Ecosystem.

[57] SOLANES, M., GONZALEZ VILLAREAL, F., (GWP-Tac), 1999. *The Dublin Principles for Water as Reflected in a Comparative Assessment of Institutional and Legal Arrangements for Integrated Water Resources Management,* GWP TAC Background Paper n.3.

[58] DUKHOVNY, V.A., SOKOLOV, V.I., 2005. *Integrated Water Resources Management-Experience and Lessons Learned for Central Asia towards the Fourth World Water Forum.* Tashkent: SIC ICWC-GWP CACENA.

- Transition from water management within administrative units towards water management according to catchments or irrigation systems (hydrological boundaries);
- Moving from sectoral water management towards a integrated cross-sectoral one, including surface water, ground water, and return water as well as integrating irrigation, domestic, and hydro-power water use;
- Transition from the authoritarian one-way principle of "top-down" water management towards a more democratic two-way principle "bottom up" (formulating water requirements and participation of water users in decision making) and "top-down" (establishing of water use limits (quotas) and support of water users);
- Participation of water users and other stakeholders in decision-making processes by setting up basin councils working with basin authorities, WUAs, and other sorts of water users organizations;
- Moving from supply water management towards water demand management (allocative efficiency) promoting the economic value of water; through this practice a more equal water supply will be ensured and wastes will be reduced.


This guideline, including the pillars and approach debated since the Dublin conference, provides an initial framework, although with some lacks concerning how to reach these changes and what the governments and policy makers need to undertake to achieve the practical implementation of the IWRM. In particular it focuses on the transition from water management authorities based on administrative boundaries to hydrological ones, which had not been discussed before, and emphasizes the bottom-up practices, in connection with the participation of the water users in the decision-making processes. As debated by several scholars, water professionals, and other stakeholders, in the process of implementation of IWRM framework a reflection on and a consideration of the following points are essential [59]:

---

[59] DUKHOVNY, V.A., SOKOLOV, V.I., 2005. "cit.".

- The political environment (governance): laws, international agreements, social conditions and priorities, political-economic systems and their interests; governance should accept the IWRM framework principles by transforming them into approved regulations and management mechanisms;
- Infrastructure control: hydraulic network for water supply, irrigation systems, and specific features of commanded areas;
- Water management participants: water management organizations, governmental and independent organizations, all social groups;
- Management mechanisms: institutional tools, economic tools (water charging/fees), management practices (top-down / bottom-up), environmental tools (for achieving the highest potential water productivity and reducing wastes);

These aspects are fundamental in order to start a national discourse at the governmental level about the IWRM and to achieve its implementation; a reflection on these points allows the comprehension of the importance of a wide spectrum of aspects, in particular the interests of the political system, the necessity of an appropriate legal framework, and the participation of all the actors, to wholly undertake reforms towards the enactment of the IWRM's goals.

### 1.3.4 The strengthening of the IWRM and its implementation

The above-mentioned issues and the general implementation path were highlighted and discussed in depth at the Second World Water Forum held in The Hague (Netherlands) in 2000. Unlike the Dublin Conference this meeting involved a wide range of stakeholders from developed and developing countries focusing on water and food security, ecosystem protection, civil society's empowerment, and transboundary river management.[60] In addition water services' privatization and public-private partnerships were debated and criticized by some of the water professionals from developing countries who considered it more relevant to focus on public access to water resources in order to reduce

---

[60] RAHAMAN, M., VARIS, O., 2005. "cit.".

poverty and promote equitable development.[61] Nevertheless, the Forum was successful as a wide debate on IWRM and for putting the framework on the political agenda. International support for the new water framework was strengthened in 2002 at the World Summit on Sustainable Development held in Johannesburg (South Africa). The Summit's Plan of Implementation recognized the IWRM as a key component in the effort to achieve sustainable development and put it at the top of the international agenda.[62] Though during the end of the 1990s the role of the GWP had not been dominant in the global water scenario, after the Second Water Forum and the Johannesburg Summit, the GWP gained the leading role in coordinating the *Framework for Action* for IWRM and for disseminating this concept throughout the world as the global water paradigm for the 2000s. Since the IWRM concept was partly criticizedin the previous years by different national and international actors because of its unclear definition and goals, in 2000 the GWP prepared and issued the following definitions in order to provide a common framework and to allay the doubts expressed:[63]

- Integrated Water Resources Management (IWRM) is a process which promotes the coordinated development and management of water, land, and related resources, in order to maximize the resultant economic and social welfare in an equitable manner without compromising the sustainability of vital ecosystems
- IWRM is a process which aims to ensure the coordinated development of water resources with a view to optimising social and economic welfare without compromising their sustainability

As mentioned by Snellen and Schrevel (2004), these definitions are the first authoritative definitions of the IWRM, and it should be noted that the management of water resources is defined, for the first time, as a process.[64]. Van der Zaag (2001), expert for GWP in South Africa, added that decision-making processes would involve the integration of the different objectives and the priorities-setting process would be carried out according to societal objectives. Moreover, he

---

[61] MERREY, D.J., et al., 2005. Integrating Livelihoods into Integrated Water Resources Management: taking the integration paradigm to its logical next step for developing countries, *Regional Environmental Change,* 5.

[62] RAHAMAN, M., VARIS, O., 2005. "cit.".

[63] GWP, TAC 4, 2000. *Integrated Water Resources Management,* Global Water Partnership, Technical Advisory Committee Paper, Stockholm, Sweden.

[64] SNELLEN, W.B., A. SCHREVEL, 2004. "cit.".

stated that spatial scales will be considered in terms of geographical variations of water availabilities and upstream-downstream interactions, as well as seasonal and regional water needs.[65] In presenting its actions for the strengthening of IWRM, the GWP assured the provision of platforms at various levels to facilitate dialogues resulting in policies and institutional changes. Furthermore, the organization claimed to provide the intellectual leadership for an integrated approach to water resources management creating the GWP toolbox: a public up-to-date knowledge centre with the tools, references, and case studies needed for implementing the IWRM.[66] In order to make the IWRM's path clearer, the GWP provided four steps to classify the implementation process of the different countries depending on their temporary results:

- Step 0: countries have not yet established an IWRM Plan at the national level.
- Step 1: countries have established an IWRM Plan approved by the government.
- Step 2: countries have successfully mainstreamed water resources management into the national development processes, with ownership at the highest levels of government
- Step 3: countries are integrating water resources management with other key sectoral processes and national priorities, achieving policy coherence.

Although the IWRM implementation path's phases were quite clear, how to use the toolbox and what exact measures and policies need to be undertaken by the governments and societies to reach those steps was less clear and incited debate. Biswas (2008) claimed that the examples provided in the so-called toolbox have never received objective scrutiny and no independent evaluation was ever made to determine if the tools were actually used and resulted inmeasurably improving water management that would not have happened otherwise.[67]

---

[65] VAN DER ZAAG, P., 2001. *Principles of Integrated Water Resources Management,* Waternet module IWRM 0.1, 1st draft, HIE-Delft and Department of Civil Engineering, University of Zimbabwe.

[66] GWP, 2006.*What is GWP?,*Stockholm, Sweden.

[67] BISWAS, A.K., 2008. "cit.".

## *1.3.5 The IWRM as the global water paradigm*

Nevertheless, starting from the 2000s all the international donors and development banks—from the World Bank to USAID and the Asian Development Bank—embraced the IWRM as the new global water resources management framework, supporting and establishing several development projects throughout the world, in particular in developing countries. This global support and acceptance of the IWRM framework since the 2000s has led to a deep reflection within academic and professional spheres. Allan (2003), recognizing that the environmental and economic phases of reflexive modernity are still in progress, argued that those steps are being supplemented by the third sub- paradigm which is based on the notion that water management and allocation is a political process; this approach is based on the IWRM.[68] According to his analysis, the IWRM demands much more than the mere recognition of the environmental and economic value of water and the planning of technical interventions. The IWRM is an intensely political process because it requires the interests and the involvement of the state (governments and hierarchies), the civil society, the NGOs, and the private sector. Allan added that this water sub-paradigm has brought forward approaches which include participation and inclusive political institutions to enable the mediation of the conflicting interests of water users and the agencies that manage water.[69] It is hence quite explicit that the interests of the society, the economy, and the environment should be simultaneously considered and debated. In the same perspective, Mollinga and Gondhalekar (2012) added that the reflections and debates in the water community after the water summits of the 1990s led to the alignment of three big ideas in the global discourse on water resources management: the ideas of market, democracy (often phrased as good governance), and sustainability, which were assembled in the IWRM framework.[70] As mentioned in the previous paragraph and above, the three sub-paradigms of reflexive modernity have affected the "North" and only a few countries of the "South" in the last decade; for these reasons the mission of several international donors and development banks have been, in the last decade, to extend the reach of the IWRM framework (including the environmental/economic/political dimensions) in order to reduce the differences and

---

[68] ALLAN, T., 2003."cit.".
[69] ALLAN, T., 2003. "cit.".
[70] MOLLINGA, P., GONDHALEKAR, D., 2012. "cit.".

contentious discourse, (Allan, 2003) in worldwide water management approaches.

### 1.3.6 The political nature of IWRM

Regarding Allan's definition of the third sub-paradigm, Mollinga (2008) claimed that the IWRM and the current water resources management discourse in general, starting from the 2000s, is inherently a political process; although until ten years ago the term "politics" was anathema in most water policy circles, the rise of the themes of governance, participation, accountability, and involvement, thrust politics into the water resources development discourse through the backdoor.[71] As the term "management" replaced the term "operation"(which was considered to involve only technical issues) in the 1970s, after the 2000s governance became a core theme, and it is not possible to separate the term "governance" from political discourse. Mollinga (2008), arguing the case, added that the current water resources management is a political process because it is based on the idea that *water control* is at the heart of water management practices and discourse, and should be conceived as a process of *politically contested resource use*. Any human intervention in the hydrological cycle affecting spatial and temporal water availability is a form of *water control*. According to him water control has three dimensions: technical - physical, organizational / managerial and socio-economical/regulatory.[72] It is questionable whether these forms of water control, which differ throughout the world, could be condensed into the IWRM and advocated through its pillars. Though Allan and Mollinga clearly stated that the IWRM is a political process, Molle (2008), agreeing with them, claimed that the promoters of the framework tried somehow to hide its political nature behind the term "participation and good governance" which is actually a political process.[73]. It was mentioned before that starting from the 2000s, the GWP organized the set of rules and practices named the IWRM toolbox, in order to facilitate the IWRM implementation throughout the world according to different physical and sociopolitical contexts.

---

[71]  MOLLINGA, P., 2008. Water, Politics and Development: Framing a Political Sociology of water resources management, *Water Alternatives,* 1 (1).

[72]  MOLLINGA, P., 2008. "cit.".

[73]  MOLLE, F., 2008. "cit.".

### *1.3.7 Combining the IWRM and the decentralization initiatives: The IMT and the Water Users Associations (WUAs)*

Since the widespread support of the IWRM concept, particularly in developing countries, has been carried out by international donors and development banks, those actors, besides the GWP toolbox, have promoted initiatives, related to the Dublin principles and more generally to the current global water resources management discourse, characterized, as argued by Mollinga (2008), by the ideas of market, democracy, and sustainability. Analysing theoretically the relations between concepts, narratives, and models, Molle (2008) has argued that models are based on particular instances of policy reforms and development interventions which embody a dimension of success and qualify as "success stories".[74] They are apparently sanctioned by experience and approved by the experts as well as powerful institutions like the United Nations, the World Bank, and others. The promoted models in the water sector which have emerged in the last decades are the Irrigation Management Transfer (IMT) and the related Water Users Associations (WUAs). Both the concept and the initiative, developed and supported by the international donors, emerged from the decentralization processes and generally from the roll-back of the state in operation, maintenance, and financial support of public infrastructures which occurred in several countries at the end of the 1980s. Several development projects have been based on concepts of the Irrigation Management Transfer (IMT). According to Ghazouani et al (2012), the IMT refers to the process that seeks the relocation of responsibilities and authority from central government entities managing irrigation schemes to non-governmental agencies, such as Water Users Associations (WUAs), or private entities.[75] Molle (2008), referring to Mexico, argued that the IMT process was part of the so-called *structural adjustment* under a neo-liberal model of economic deregulation, downsizing of the public sector, reduction of public expenditures, and the reconfiguration of the public administration responsibilities among national and local levels.[76] Ghazouani et al. (2012) stressed that the donors shrouded their projects in participatory rhetoric, promoting the IMT as a model featured by the co-managing of the infrastructures by the water users, with the will to lead to a bottom-up sense of ownership (of infrastructures and organizations).[77] Although most scholars, both from academic and professional

---

[74] MOLLE, F., 2008. "cit.".

[75] GHAZOUANI et al., 2012. "cit.".

[76] MOLLE, F., 2008. "cit.".

[77] GHAZOUANI et al., 2012. "cit.".

spheres, stated that this process might lead to an increase in participation and inclusion in management processes by the water users, Yakubov and Ul-Hassan (2007) discussed that participatory management may also lead to a discrepancy between the marginalized poor and the powerful groups, appropriating the reforms' benefits.[78] Nevertheless, the IMT aims at deep changes in the relations between the state agencies (for instance, the basin agencies) and the water users, providing the knowledge to set-up organizations according to a participatory approach and bottom-up practices as well as to self-maintain the irrigation systems at the local level. It is questionable whether this approach could be implementable and could lead to benefits in all physical environments and sociopolitical systems. It was also mentioned that the IMT would lead to benefits in governmental budgets, reducing governmental spending in irrigation systems maintenance; Wegerich (2006) claimed that several governments affected by the financial crisis, particularly those of transitional countries, such as the former Soviet Union and Eastern Europe, could no longer maintain the subsidies for large irrigation schemes.

### 1.3.8 The WUAs as a worldwide example of local level water management

The worldwide example of the IMT, particularly widespread in transitional and developing countries, has been the establishment of the Water Users Associations (WUAs) or Water Users Organizations (WUOs) or Water Users Unions (WUUs). Salman (1997) defined them as groups of farmers, usually comprised of one hydraulic unit, command or irrigation district, organized as a non-profit organization for the purpose of managing parts or whole irrigation systems, based on a self-organized and participatory approach.[79] The primary objective of a WUA is to achieve optimum utilization of available water in a sustainable way, endowing the users with a major role in the management decisions over water in their hydraulic unit. Focusing on territorial size and members, the established WUAs throughout the world range from 200 to 300 hectares to more than 5000 hectares, and from 10 to 20 people to 1000 to 2000 people. Reflecting on WUAs and its features, Hunt (1989) suggested that for a WUA to be successful in their responsibilities, allocation, accounting, and maintenance, they should be small in size and in the number of members.[80] According to Ghazouani et al

[78] YAKUBOV, UL-HASSAN, 2007."cit.".

[79] SALMAN, M.A (1997) *The Legal Framework for Water Users' Associations-A comparative study.*World Bank Technical Paper n.360.

[80] HUNT, R.C., 1989. Appropriate Social Organizations? Water Users Associations in Bureaucratic

(2012), the established WUA should be structured on three domains of responsibility: water management, maintenance, and financial management. Through those domains the farmers should be able to participate in decision-making processes, planning water allocation schedules, maintaining the water facilities and the outlets, and collecting the fees for WUAs' financial budget.[81] Depending on several sociopolitical factors, different WUAs' variants have been established worldwide—in some cases including a formal governance council or only informal meetings, in other cases managing all irrigation schemes or just the tertiary level. According to Salman (1997), an institutional framework and governmental support is necessary for proper WUAs performance; he mentions the enabling law, the bylaws of the WUAs and the transfer agreement between the irrigation agency (state agency or department) and the WUAs. Regarding the WUAs' establishment, its performance, and sustainability, Wegerich (2006), reviewing the statement of several scholars (Huppert 2001; Jordan 2001; Meinzendick 1994; and others), claimed that the WUAs' performance is directly influenced by external and internal factors, complementary among each other. Focusing on the external factors, he mentions the Physical & Technical, Policy & Governance, and Social & Economic; while internal factors include the bylaws, the structural organization, the membership criteria and the group dynamics. Therefore, it might be questioned whether those models, mostly developed by the Western water community within the support of the wider IWRM framework, could be efficiently implemented in developing countries, particularly in those still featured by a state-centralized approach in natural resource management. Ghazouani et al. (2012) recently analysed the establishment and related performance of the WUAs in several parts of the developing world, from the Middle East to Northern Africa as well as in the former Soviet Union; according to them, the evidence has shown that the IMT combined a mixture of pragmatic material needs (less state expenditures) with ideological fervour, and that several WUAs lack participation of the water users and are strongly influenced by the water bureaucracies.[82]

---

Canal Irrigation System, *Spring,* vol. 48, n.1.
[81] GHAZOUANI et al., 2012. "cit.".
[82] GHAZOUANI et al. (2012). "cit.".

## 1.4 THE DEBATE AROUND THE IWRM

After the first years of collective fascination and enchantment which followed the world environmental and water summits, since the mid 2000s the IWRM framework and related models have started to be debated and questioned within academic circles, specifically their implementation in developing countries. The focus of the debate has been on a wide spectrum of issues ranging from the definition of the IWRM, its procedures and aims, and on how to structure its implementation process in terms of institutional / political and economic changes throughout the world. This uncertainty about the validity of the IWRM as a global paradigm, was partly due to the superficiality and unclearness of the homonymous toolbox, the implementation issues which emerged in different countries, and the lack of quick improvement which had been expected. Furthermore, the single pillars of the framework were questioned — for instance, whether the basin management unit would be the best structure to manage water; or how to manage, institutionally and practically, integration of water use; and whether the discourse on the economic value of water could be considered in countries affected by poverty and lack of water access.

### 1.4.1 Debating the IWRM definition

The last years' debate around the validity and success of the IWRM framework has shown different and contrasting positions among different scholars, ranging from a severely critical approach to recognizing the need to reconsider parts of its pillars; others totally supported, the framework.[83] Focusing on its definition, Jonker argued that the conceptual base of the IWRM is not clear and that the GWP definition does not provide the theoretical clarity required for practitioners to achieve successful implementation; he added that the definition is a bit elusive and does not explain the terms — for instance, the maximization or coordinated development of water, included in the definition.[84] Allan (2003) claimed that the IWRM approach can be only deployed if two relevant conditions were taken into account: firstly, IWRM must be seen as primarily a political process in terms of getting policies in place; then, he added that although this nature was not mentioned in its definition, it is clear that the framework is reshaped in several cases by local political imperatives. According to him the Integrated Water

---

[83] JONKER, L. (not Ment.). Integrated Water Resources Management: the theory-praxis-nexus, IWRM programm, University of the Western Cape, South Africa, Working Paper.

[84] JONKER, L. (not ment.). "cit.".

Resources Management should be renamed IWRAM—Integrated Water Resources Management and Allocation—because allocation and re-allocation are unavoidable in water policy and management; he added that these processes are always contentious and political.[85] Nevertheless, regarding Allan's statements, Jonker argued that just adding allocation to the GWP definition still does not render it capable of assisting and guiding implementation.[86] Merrey et al. (2005) also criticized the GWP definition of the IWRM, from a different perspective, claiming that it is too narrow and elusive; they stressed that it is evident that it was considered and designed according to the *western* rationale. They claimed that the IWRM does not give enough consideration to the empowerment of poor communities, poverty reduction, and the improvement of livelihoods, focusing instead on second generation issues like cost recovery, re-allocation of water use, and environmental protection.[87] Merrey et al. (2005) promoted a new definition of IWRM, focusing on poverty reduction, but somehow still wider and more narrow: IWRM should address the promotion of human welfare, especially the reduction of poverty and the encouragement of better livelihoods and balanced economic growth, through effective democratic development and management of water and others natural resources at community and national levels in a framework that is equitable, sustainable, transparent, and far as possible to conserve vital ecosystems.[88] According to Van der Zaag (2005), who works in South Africa and therefore is directly facing poverty issues, IWRM is a framework based on a balance in using water for achieving social aims, economic development, and ecological protection; he added that the framework is a must, all countries and institutions have to embrace it, and that it is currently the best practice in water management, which in the future should inspire new generations of water experts and professionals.[89] According to an opposite idea of the IWRM, Biswas (2008) expresses severe criticism of its definition from multiple perspectives; he stressed that Integrated Water Management is based on the rediscovery of the old concept of "integration" formulated in the 1960s which

---

[85] ALLAN, T., 2003."cit.".

[86] JONKER, L., (not.Ment.). "cit.".

[87] MERREY et al. (2005). Integrating Livelihoods into the Integrated Water Resource Management: taking the integration paradigm to its logical next step for developing countries, *Regional Environmental Change,* 5.

[88] MERREY et. Al. (2005). "cit.".

[89] VAN DER ZAAG, P., (2005). IWRM: relevant concept or irrelevant buzzword? A capacity building and research agenda for Southern Africa, *Physics and Chemistry of the Earth,* 30.

could not be implemented in the previous decades because of its elusive approach and contrasts inside the framework. Biswas stresses that the old concept was reformulated by including in its definition fashionable and trendy words like sustainability and participation, providing an amorphous, wide and fuzzy concept which could not help water professionals to effectively solve the different and contrasting water issues throughout the world.[90] He also mentions that IWRM was formulated only by water professionals from the developed countries without an effective involvement of experts from the developing ones. Biswas (2008) deeply analyses the IWRM's GWP definition posing several questions: from one side he criticized the term "maximize", arguing that it is not clear what specific parameters should be maximized and how and who will have to select them; from the other side, he critically focused on the term "integration". He identifies 41 sets of issues (ranging from water quantity and quality, urban and rural issues to government and NGOs, industrial and hydropower water use, and others) which different institutions considered to be the issues which should be integrated in the main framework of IWRM but simply cannot be achieved because of internal contradictions, and too generally described goals without effective guidelines for their implementation. In addition,he argues that since in the current complex world several issues like energy, water, agriculture, and rural development are interrelated and interdependent, the emphasis only on water, despite its challenging institutional and managerial integration is not wide or broad enough.[91]

### 1.4.2 Debating the Integration pillar

Although the emphasis on integration is at the core of the IWRM, it is not clear how this integration among water uses and water demands could be effectively implemented from an institutional and organizational point of view and no guidelines were provided. Regarding these issues, Allan (2003) claimed that integration is also a political process as all those who have attempted to take an interdisciplinary approach know, but have not officially mentioned.[92] The integration of different ministries and institutions — for instance, combining the ministry of water with energy, environment or industry ministries — requires relevant institutional changes supported by a strong governmental will; furthermore, it would be featured by different issues in the different countries through-

---

[90] BISWAS, A.K., 2008. "cit.".
[91] BISWAS, A.K., 2008. "cit.".
[92] ALLAN, T., 2003. "cit.".

out the world. Merrey et al. (2005) do not focus on the challenging process of institutional integration but at the same time they argue that integration will be the key concept for the future and that it requires a more holistic, participatory, and interactive approach among scientists, professionals, and politicians. They suggest an integration across scales, components, stakeholders, and disciplines.[93] Hence, according to them, the IWRM concept should be included in a wider and broader INRM, Integrated Natural Resources Management, focusing on theimprovement of livelihoods. Tools for operationalizing INRM are provided: systems modelling, participatory action research with stakeholders, multiscale databases, impact assessment, and GIS.[94] Nevertheless, it is questionable how this INRM, which needs a mature interdependent approach and also advanced informatics systems could be operational according to different sociopolitical systems.

### *1.4.3 Debating the commodification of water*

Merrey et al. (2005) have not been the only ones who have criticized the weak emphasis of IWRM on poverty reduction and livelihoods' improvement; Rahaman and Varis (2005) claim that an IWRM oriented to second generation water problems will not adequately help the poor communities of developing countries in achieving water security. Furthermore, they stress that the privatization of water services and the idea of "water as an economic good" should be slightly rethought; they claimed that privatization of the marketable aspects of water may result in a single purpose planning and management, which raises a question of open information channels and transparency. In addition, in some developing countries, the poorest for instance, a question remains whether applying full cost recovery would be ethical or practical.[95] Thus, although the application of economic principles to the allocation of water is advantageous in terms of providing efficiency, it should not be treated as a market-oriented commodity when it comes to domestic use for very basic needs, particularly in extremely poor communities. Rahaman and Varis also support the idea that environmental protection and sustainability cannot be at the top of the agenda since socio-economic development is more urgent. Allan (2003), stating the distances of the water process in places among the "North" and the "South", claims that insisting, throughout the IWRM support, on preaching about the

---

[93] MERREY et al, 2005. "cit.".
[94] MERRET et al, 2005. "cit.".
[95] RAHAMAN M.M., VARIS, O., 2005. Integrated Water Resources Management: evolutions, prospects and future challenges, *Sustainability, Science, Practice and Policy,* vol.1, issue 1.

environmental and economic value of water will have little impact in many of the developing countries' communities affected by water scarcity.[96] According to a different perspective, Van der Zaag (2004) claims that the economic value of water is still a debated issue for policy makers and governments in the "South" and that Southern Africa needs scholars and water managers who have a critical understanding of the limitations and opportunities that the market and private sector have to offer.[97]

### 1.4.4 Debating the hydrographic unit in water management

Specifically related to the territorial features, another debated pillar of the IWRM is water management according to basin units. Ostrom (1990) mentions "clearly defined boundaries" as the first principle in her recommendations on natural resources management.[98] Even engineers tend to naturalize irrigation systems and their boundaries. Mollinga (2007) explains that system-level and hydraulic boundaries of irrigation infrastructures are represented as being the natural management unit for irrigation systems, according to the ecosystems.[99] The choice of the river basin as the ideal unit for IWRM has been debated over by several international actors like the European Union throughout the EU water framework directive and the EU water initiative, presented at the Water Forum in Johannesburg in 2002. Moreover, the Global Water Partnership has strongly promoted the basin unit since water flows according to natural characteristics without considering administrative boundaries.[100]. Nevertheless, as Graefe (2011) stressed, the high complexity of the present water management practices and the connectivity of river basins through water transfers show that catchment areas are not the unit of water management in large parts of the world.[101] In fact, despite the strong support for the basin unit by the IWRM and international donors, in a large part of the world, in particular in those areas characterized by the *hydraulic mission*, until a short time ago, water was still managed according to administrative boundaries. Furthermore, Merrey et al. (2005) stated that creating

---

[96] ALLAN, T., 2003. "cit.".

[97] VAN DER ZAAG,, P. , 2005. Integrated Water Resources Management: Relevant concept or irrelevant buzzword? A capacity building and research agenda for Southern Africa, *Physics and Chemistry of the Earth,* 30.

[98] OSTROM, E.. 1990. *Governing the Commons: The Evolution of Institutions for Collective Action.* Cambridge University Press.

[99] MOLLINGA, P., 2007. "cit.".

[100] GRAEFE, O., 2011. River basins as new environmental regions? The depolitization of water management, *Procedia Social and Behavioral Sciences,* 14.

[101] GRAEFE, O., 2011. "cit.".

new water institutions based on hydrographic boundaries, in particular in parts of developing countries, requires a challenging and expensive political institutional change, which does not ensure an improvement in water management practices.[102] It is therefore questionable whether managing water according to the catchment unit, as promoted by the IWRM, is the best practice throughout the world without considering the different managements and institutional systems as well as the political context. In relation with these issues, Graefe (2011) also claimed that the choice of the river basin as a planning unit should be questioned due to the increasing water transfers between catchment areas.[103]

### 1.4.5 Debating the worldwide implementation of the IWRM

Although the IWRM pillars have been deeply questioned and discussed during the last decade by water professionals, academia and donor members, the main question, as mentioned before, has been and still is whether the IWRM, according to the present definition, pillars and guidelines, can be efficiently implemented. A further and related question asks whether the framework would be implementable throughout the world, despite its physical, environmental, socio-political, and economic differences, leading to benefits and improvements in water management. Jonker, reviewing the literature on the debate about the implementation of IWRM, stated that several positions among scholars have emerged; according to his review, on the one hand, Allan (2003) and Merrey (2005) claim that the concept should be partly reworked to be implementable, and on the other hand, Rahaman and Varis (2005) and Van der Zaag (2004) stress that only certain issues should be further addressed.[104] In contrast, Biswas (2008) takes a deeply critical and extreme position towards the IWRM and its implementation. Allan (2003), before focusing on the implementation process, stressed that the water policies significantly differ between the "North" and the "South" and also throughout the countries of these two "realities"; therefore, it is challenging to implement the same framework if considering considerably different and contrasting water scenarios. Nevertheless, he claimed that implementation is easier in the "North" countries, since the paradigm was designed and created by the professionals and policy makers belonging to this part of the world, and the political and economic pre-conditions are more favorable in comparison with most of the global "South" countries. Considering the IWRM

---

[102] MERREY et al, 2005. "cit.".
[103] GRAEFE, O., 2011. "cit.".
[104] JONKER, L., (no ment.). "cit.".

implementation to be a political process, Allan (2003) states that it requires the involvement of all the actors—that is, the government, the private sector, the NGOs, and the civil society.[105] Therefore, without the active involvement of all these actors the IWRM is not implementable; furthermore, he claims that all the stakeholders must know and want the promoted reforms. *Knowing about, having,wanting, operating and effectively operating (KHWOE)* the water reforms is the milestone for changing the sociopolitical priorities which are the requirements for the IWRM implementation.[106] It is already questionable whether in several countries, particularly in the developing world, the involvement of all the stakeholders and the subsequent KHWOE process could be possible. Strengthening and widening the debate, according to Merrey et al. (2005) the IWRM, to be implementable throughout the world (in particular in developing countries), should focus more on the improvement of livelihoods and poverty reduction. In addition, they focus on two main issues: scale and governance. Regarding the scale, the river basin unit is an higher level whether considering the livelihoods needs and issues, and therefore the IWRM should focus more on communities and the local levels. Focusing on governance, Merrey et al. (2005) stressed that significant problems in setting up governance structures in some of the developing countries have emerged, and, therefore, empowerment of the local communities, knowledge, and organizational/technical assistance by the developing agencies are required to put IWRM into practice.[107] Rahaman and Varis (2005) added that it is necessary to reduce the gap between theory and practice to make IWRM implementable; the toolboxes provided by the GWP have not been effective in supporting the implementation path.[108] Therefore, they also claim that deep reflection on the main pillars of the framework is necessary to strengthen its implementation. Rahaman and Varis (2005) specifically stress that the water professionals should reflect and partly reconsider the theme of privatization and the idea of "water as a market commodity", as well as the concept of integration and environmental sustainability, since these notions can lead to different consequences and issues depending on the part of the world and the particular countries that are involved.[109] Their priorities and relative importance vary enormously from one place to another, as do the challenges to reform

---

[105] ALLAN, T., (2003). "cit.".

[106] ALLAN, T., (2003). "cit.".

[107] MERREY et al. (2005). "cit.".

[108] RAHAMAN, T., VARIS, O., 2005. "cit.".

[109] RAHAMAN, S., VARIS, O., 2005. "cit.".

the water sector. In addition, they claimed, according to Allan's perspective, the IWRM cannot be universal because of the different priorities, environments, and role of water in the different countries throughout the world. Hence, the IWRM could be reduced to an idealistic buzzword if water professionals fail to overcome the above-mentioned reconsiderations of the framework. In contrast to these authors, Van der Zaag (2004), being directly involved with the implementation of IWRM in Southern Africa, stated that the framework is currently the best practice in water management and, therefore, it should be supported by all the stakeholders.[110] Nevertheless, he argued that the IWRM can be reinforced focusing on the following themes: the institutional dimension (that is, the capacities of institutions to integrate the new organizations based on hydrographic boundaries with former ones based on administrative units), the decision-making processes (water professionals have to better support the participation of the stakeholders), and the upstream-downstream issues (which underscore the necessity for strengthening cooperation both at national and inter-state levels). Van der Zaag added that throughout the IWRM, the relations between the government and the citizens could be redefined and reinforced; he provided the good example of its implementation in Southern Africa.[111] In contrast, Biswas (2008) takes an extreme and vehemently critical position to the IWRM implementation which leads subsequent critics to his idea undertaken by the policymakers of the global water community. After the critics to the GWP's definition (mentioned in the previous paragraph), Biswas argues that the implementation toolbox is extremely unclear, superficial and insufficient, and the guidelines provided to put IWRM into practice useless. In addition he stressed that most of the developing countries which agreed to implement the IWRM undertook this decision in order to get money and international visibility, and finally they have not effectively changed their water sector without any monition expressed by the developing agencies. Moreover, he questioned whether IWRM could really lead to benefits in all the regions without leading to unexpected disputes or issues —for instance, as the consequences of the integration of different ministries and institutions.[112] Biswas, according to the ideas also pointed out by Allan (2003) and Rahaman and Varis (2005), questioned how a framework can be valid for the whole world when each place and situation presents different and complex con-

---

[110] VAN DER ZAAG, P., 2004. "cit.".
[111] VAN DER ZAAG, P., 2004. "cit.".
[112] BISWAS, A.K., 2008. "cit.".

texts and issues, different physical environments, water availability, and related demands, different roles of water in state economies, different perspectives on economy and development, different cultures and connected norms, and finally different and often contrasting policies and governments, and related ideas of decision-making processes, governance, and participation,[113] To support his analysis, he claimed that throughout the world the implementation level of IWRM, on a scale from 0 to 100 according to different parameters, still has not reached more than 30%; therefore, he predicts in less than a decade this water paradigm will fail and subsequently fall down. Nevertheless, Molle (2007), sharing Biswas's ideas regarding the loans, money, and incentives gained by the countries which are going to reform their water sector towards the IWRM, argued that the paradigm maintenance is really an ideological point which aims to spread economic and political ideas throughout the world — for instance, nowadays, the efficiency of privatization, bank investments, and state action reduction.[114] Therefore, a shift in paradigms is a genuinely challenging process, being under the control of strong governments, international actors, and developing agencies. In the following chapters the IWRM framework, its pillars, and in particular its implementation process will be further analyzed in order to go more in depth and enrich the above-presented debate. Although the position of Biswas (2008) on the IWRM is exceptionally critical and quite extreme, specifically in presenting the worldwide implementation scale, he is the author who has most questioned and stressed the possibility for the IWRM to be implementable worldwide, in particular in developing countries. Biswas in particular focuses on how a framework that has been discussed and designed in Europe by Western water professionals and according to their related sociopolitical environment, can fit and lead to environmental or social benefits throughout the world, particularly in countries that really differ, concerning political-economic systems, from the Western ones. Despite the fact that the Global Water Partnership and other donors have stated that the IWRM would lead to multi-perspective benefits throughout the world, and the toolbox provided would help governments and policymakers to implement it, it is questionable whether the different governments would consider the whole framework or just parts of it which better fit their local political and economic systems. In fact, since the IWRM implementation is a real political process , as argued by Allan and others, it would be rele-

---

[113] BISWAS, A.K., 2008. "cit.".
[114] MOLLE, F., 2008."cit.".

vant to understand how the local stakeholders, including the government, civil society, and private actors can influence the implementation path according to a national perspective. Furthermore, it seems essential to analyze the relations among the stakeholders and the different powers in the inner decision-making processes in order to understand the dynamics which affect and influence the choices of which specific IWRM pillars are to be implemented. Concerning the evidence analyzed and highlighted by other authors, such as Molle and Mollinga (2008) and Van der Zaag (2004), some countries decided to focus on the partici-patory approach, increasing participation in the decision-making processes, without considering the hydrographization of the water authorities; other coun-tries supported this process while leaving out the introduction of water fees and keeping a top-down approach, thus limiting the action of the civil society.[115] Therefore, it seems clear that these choices are related to the countries' aims and strategies, and are related to the national political-economic trajectories. More-over, the analysis of these strategies allows the comprehension of the power of the different stakeholders; even though Allan (2003) claims that the IWRM im-plementation requires the active involvement of all the stakeholders and the re-lated KHWOE process, a partial implementation shows the priorities and the strategies of the most powerful actors have been accomplished. Therefore, aim-ing to explain and understand in depth the current IWRM implementation proc-ess, the selection of the pillars and related national reforms were considered. Mollinga (2007) claimed that although the water reforms are formulated and en-acted at the national level, the effective implementation and related issues emerge at the basin-local level; therefore, this scale was chosen for analysis.[116]

### 1.4.6 Analyzing the logics of the IWRM

As it was widely discussed in the previous paragraphs, at basin-local level the basin agencies and the Water Users Associations (WUAs) have been the models supported and widely adopted since the 1990s by the IWRM framework and the Irrigation Management Transfer (IMT). Therefore, in order to answer to the re-search questions posed here, the focus is on the water authorities at the basin level, and on the Water Users Associations (and related water users and farmers) and former district water departments at the local level. Of these authorities the

---

[115] MOLLE, F., MOLLINGA, P., 2008. "cit.".
   VAN DER ZAAG, P., 2004. "cit.".
[116] MOLLINGA, P., 2007. "cit".

following aspects, in relation with the IWRM pillars, were considered and analyzed:

- the institutional structure (law, institutional status) and the organizational characteristics (members, operational features) in connection with the national reforms
- the boundaries of the authorities' territories (in connection with the supported hydrographization of the water authorities)
- the integration and the participation of the stakeholders — water users and members — according to a participatory approach in the decision-making processes (in connection with the supported bottom-up practices and governance)
- water fees collection (in connection with the supported economic value of water)

Moreover, specifically concerning the water users associations, the personal position and opinions of the water users have been highlighted in order to understand the effective performance of the associations and the related success of the Irrigation Management Transfer. For this analysis, a physical description of the territories and related canals networks and water infrastructure was undertaken in order to have a complete overview of the territorial evolution of thecase studies. The analysis of the afore mentioned aspects of the former and newly established water authorities at the basin and local level has allowed firstly a reflection on the implementation level of the national reforms in the local context. Then, in a wider perspective, it has allowed a deep reflection and the possibility to build a strong argument on the current IWRM framework implementation and national trajectories over water policies and related issues in developing countries, specifically in the Central Asian region. The comparison of two case studies in two different countries gives the possibility to highlight the potential differences and similarities regarding the IWRM's perspective throughout Central Asia. Therefore, this discussion fills the apparent void in the research of these topics in the Central Asian region and, in addition, as a main aim, strengthens and enriches the current above-mentioned scientific debate regarding the analysis of the IWRM's implementation procedures throughout the world.

# 2. COMPARATIVE METHODS IN WATER STUDIES: THE METHODOLOGICAL APPROACH

## 2.1 A COMPARATIVE APPROACH IN SOCIAL AND WATER STUDIES

In the previous chapter the transition and decentralization processes that affected the water sector in the Central Asian countries after the collapse of the Soviet Union and during the 1990s were analysed; the evidence shows that those processes followed different paths according to the different countries' s situations, due to sociopolitical and economic strategies as well as to the relevant international donors' actions, and occurred in shorter or longer periods. As mentioned in the introductory section, Uzbekistan and Kazakhstan were selected as research countries according to several features ranging from territorial to political-economical ones. According to Mollinga (2012), since the measures and the water reforms are decided and issued at the national level by the governments involved, the effective practice and their implementation is strongly related to the territories and basins' political authorities.[117] Therefore, within Uzbekistan and Kazakhstan two basins were selected and deeply analysed in order to understand the trajectories of water resources management reforms: Middle Zeravshan valley, located in central-eastern Uzbekistan, and Arys valley, which lies in south-western Kazakhstan. Focusing on the research methodology, it is important to point out why a comparative approach has been chosen to conduct the analysis and why it could be defined as a suitable method within the main field of social water studies with a human/political geographical perspective. Comparative research aims to identify significant differences (qualitative differences in structural and operational configurations) between different contexts or situations; thereby comparative research can potentially help to define "relevance domains" for specific policy interventions and consequences. It may be the appropriate method to analyse the implications of institutional and political transitional context in different territories, highlighting the connected strategies that followed these common changes; for instance, the different paths pursued by the Central Asian countries after the USSR's dismantling or the water strategies that occurred in neighbouring Pakistan and India after the British Empire's collapse. Comparative analysis as a research methodology has a long history in the social

---

[117] MOLLINGA, P., GONDHALEKAR, D. 2012. "cit.".

sciences. Since the huge political and economic transformations that occurred in Europe in the last centuries, comparative analysis has been used by several scientists to find new ways of understanding these large-scale phenomena, for instance comparing national and regional dynamics and related issues.[118] Durkheim (1982) argues that all sociological research is in fact comparative since all social phenomena are held to be typical, representative or unique, all of which imply some sorts of comparison.[119] According to Ragin (1987), all empirical social research — for instance historical, geographical, or anthropological ones — involves comparisons of some sort, of real or hypothetical cases; in addition he states that the holistic approach of comparativists encourages structured, focused comparisons and a small number of case studies (for instance, countries or villages) which enable in-depth analysis and help cases to remain in the foreground.[120] Bailey added that any descriptive effort, any typology or classification, involves comparisons. According to Dogan and Pelassy (1990), comparative social and political research is generally defined in two ways: either on the basis of its supposed core subject, which is almost always defined at the level of political-social systems, or by means of descriptive features that claim to enhance knowledge about policies and society as a process.[121] These descriptions are generally considered to differentiate the comparative approach from others within social and political sciences; moreover, they must be elaborated and accomplished through a theoretical framework and a research strategy. According to Keman (1993), in comparative analysis, theory and methods are always mutually interdependent.[122] A first and vital step in the process is to reflect on the relationships between the cases under review and the variables employed in the analysis: in general, the more cases that are compared, the less variables are often available. According to the current debate on comparative analysis in social sciences, when focusing on case studies, five options are available:

---

[118] MAHONEY, J., RUESCHEMEYER, D., 2003. *Comparative Historical Analysis in the Social Sciences,* Cambridge: Cambridge University Press.

[119] DURKHEIM, E. 1982.*The Rule of Sociological methods,* (W.D. Halls Trans), New York: the Free Press.

[120] RAGIN, C.C., 1987. *The Comparative Method: Moving beyond the qualitative and quantitative strategies,* University of California Press.

[121] DOGAN, M., PELASSY, D., 1990.*How to Compare Nations: Strategies in Comparative Policies,* Chatman House Publisher.

[122] KEMAN, H., 1993. *Comparative Policies: new Directions in Theory and Methods,* Amsterdam, VU Press.

- The single case study (either a country/region or an event)
- The single case study over time (historical study or time series analysis)
- Two or more cases at a few time intervals
- All the case studies which are relevant regarding the research question
- All relevant cases across time and space

Although a single case study cannot be considered as purely comparative, it has been often used for reasons of validation post hoc to inspect whether or not the general results of a comparative analysis hold up in a more detailed analysis or to study a deviant case in a theoretical framework.[123] The single case study over time is often used in analysis focusing on a country's history with a specific focus derived from the research question; examples of such studies can include the dynamics featuring a particular political or economical process. The third option is seen as a focused comparison between countries or regions affected by the same ongoing processes which require steady temporal analysis.[124] The fourth option is the most prevalent one used in comparative analysis: it concerns those case studies that have more similarities than differences, which can be analysed and compared through a specific research framework; examples of such an approach can be the comparison of sociopolitical or environmental issues in different regions contextualised in a homogeneous environment.[125] The last comparative approach is strongly debated because on the one hand, the number of case studies is indeed maximized, but on the other hand, the time variable can be considered to be constant for all the analysed cases, keeping the approach appropriate. Since the 1970s the writing and discussions on comparative methods have significantly increased; therefore Oyen, at the beginning of the 1990s, stated that the growing internationalization and the export and import of cultural, social, and economic manifestations across national borders increased the call for comparative analysis.[126] In the last decades this research approach has been used in various disciplines — for example, in political sciences, economics, cultural, and anthropological studies, urban planning as well as water resources management. However in the course of the development of the comparative

---

[123] LANDMAN, T. 2000. *Issues and Methods in Comparative Policies, an introduction,* Routledge, London and New York.

[124] RAGIN, C., 1991. *Issues and Alternatives in Comparative Social Research,* E.J. Brill.

[125] LANDMAN, T. 2000. "cit.".

[126] OYEN, E. (ed.), 1990. *Comparative Methodology: Theory and Practice in International Social Research,* International Sociological Association, Sage pubblications.

methodology, several issues of contention between different methodological approaches have developed; social scientists have long remained polarized over whether to employ qualitative or quantitative methods. Furthermore, several scholars disagree on whether to use a large or small number of case studies (the so-called "small N/large N" debate).[127] Scholars supporting the qualitative approach underline the importance of obtaining in-depth aspects of the cases through an analysis which cannot be obtained through quantitative data analysis techniques. Ragin (1991) claims that comparative social science tends to ask questions about empirically defined, large-scale social issues and processes which are difficult to prove using quantitative methods; furthermore, comparative social studies often choose as social units case studies, focusing on their inner relations and features, which are difficult to analyse through a quantitative approach.[128] In addition, focusing, for instance, on processes analysis, the qualitative methods allow scholars to gain an in-depth understanding of social or political issues rather than quantitative approaches. On the other hand, this methodology advocates a common language of science whereby the data explains itself — for instance, numbers or statistical results; furthermore, critics argue that the large number of cases necessary for statistical analysis reduces the level of detail of the analysis in each case study.[129] Regarding the debate on the number of case studies, proponents of the small N method claim that only focusing on a limited number of cases allows for in-depth analysis. According to Ragin (1987), comparative social science has always maintained a vigorous small-N approach devoted to in-depth knowledge of particular spaces and times; furthermore, researchers using this method often combine causal analysis, interpretive ones, and concept formation.[130] In addition, small-N method scholars use flexible analytical frames that can be modified in the course of the study when knowledge emerges, making the research path less rigorous when compared to the quantitative large-N approach. Tilly (1984) also argues that comparative studies of large structures and complex processes produce more intellectual feedback when researchers examine a relative small number of case studies.[131]

---

[127] MOLLINGA, P., GONDHALEKAR, D. 2012. "cit.".

[128] RAGIN, C.C., 1987. "cit.".

[129] LANDMAN, T., 2003."cit.".

[130] RAGIN C.C., 1997. Turning the Tables: How case-oriented research challenges variable-oriented research, *Comparative Social Research,* 16.

[131] TILLY, C., 1984. *Big structures, Large Processes, Huge Comparisons,* New York, Russell Sage Foundation.

Concerning the large-N method, Ragin (1997) points out that the practical problems of this approach are how to define and delineate the classes of cases relevant to a particular investigation, how to study the causes of outcomes which are uniform across selected cases, how to study multiple paths to the same outcome, and how to account for nonconforming cases.[132] Despite these methodological differences, several scholars use a combination of different approaches, depending on the aims of the research and on the particular phase of the research path. Although the analysed debate has focused on comparative methods in social sciences as a whole, since the 1970s this methodology has been very often used in the specific field of water studies. Though in the past, water issues have been analysed through a purely engineering or natural approach, since the planned development of water resources fell into crisis, a social approach analysing notions like community development, sustainability, and participation started to emerge. According to Mollinga, in the last decades a large part of water research has been funded in close connection with international and regional development programmes, combining together technical subjects and social sciences; furthermore, some of the independent social science academic research on water resources have taken these development efforts as its subject matter, often comparing different projects.[133] In water studies much qualitative research is based on the idea of "comparison by contrasts" where the cases are chosen by researchers for the purpose of illustrating and analysing a particular issue or on the "Most Similar Case Design (MSCD)", based on Mill's method of difference. Within this method, the respective countries or regions are selected based on a number of shared features, hence the differences that explain sociopolitical or environmental outcomes can be pointed out.[134] Both the approaches, seeking out similarities or differences, enable the deduction of policy or the scrutiny of social implications; in the last decades they were both used for focusing on maximum differences — for instance, comparing traditional irrigation in dry and wet regions — or on specific similarities, such as analysing water demand in two different irrigation schemes in the same physical/political context in Brazil.[135] Through these two different but complementary approaches, several topics in the main framework of water resources management have been analysed: Irrigation Management Transfer (IMT), transboundaries river issues, water laws and poli-

---

[132] RAGIN C.C., 1997. "cit.".

[133] MOLLINGA, P. , GONDHALEKAR, N. 2012. "cit.".

[134] SEHRING, J., 2007. "cit.".

[135] GERTZ (1972) and DE NYS (2004) in MOLLINGA, P. , GONDHALEKAR, N. 2012. "cit.".

cies, the WUAs' performance, and other topics. Regarding the validity and the effectiveness on water studies, Wescoat (2003) states that the most effective studies are those driven by immediate water problems; comparative analyses become more practical when they focus on water management successes and failures for their potential relevance beyond the places and the times where they have been observed.[136] According to this trend, Sehring's research dissertation (2007) focuses on the politics of water institutional reform that followed the USSR collapse, making a comparative analysis of the recent institutional reform path carried out by Kyrgyzstan and Tajikistan.[137] In the last decade a new methodological approach, Socio-Technical Analysis, has been adopted by Mollinga in order to analyse water management and related issuesin depth. This research method analyses, through surveys and data collection, the linkages among the three dimensions of water management, or specifically of water control: physical control, organizational, and sociopolitical/economic aspects.[138] According to Mollinga, the first dimension means the physical/technical infrastructures of the irrigation schemes; the second the organizational control, institutions, and management; while the third means the water laws and policies and the social and governance structure.[139] The Socio-Technical analysis approach has been pertinent in the Central Asian region analysis where water management in the last decades shifted from a purely centralized and technical issue to a widely debated and contested sociopolitical endeavour. Abdullaev and Mollinga, in a recent research paper (2010) analysed, applying Socio-Technical analysis, the issues and dynamics related to the establishment of the WUAs and their performance in Uzbekistan (Khorezm province).[140] Having presented and discussed the methodological framework of comparative analysis, in the next paragraph the research approach will be outlined.

---

[136] WESCOAT, J.L., 2003. Water Resources. In: Gaile G.L. and C.J. Willmott (eds.) 2003. *Geography in America at the Dawn of the 21st Century,* Oxford University Press.

[137] SEHRING J. 2007. "cit.".

[138] ABDULLAEV, I., MOLLINGA, P., 2010. "cit.".

[139] MOLLINGA, P. , 2008. "cit.".

[140] ABDULLAEV, I. , MOLLINGA, P., 2010. "cit.".

## 2.2 THE METHODOLOGICAL FRAMEWORK: AN IN-DEPTH COMPARATIVE ANALYSIS AT THE BASIN / LOCAL LEVEL

In the previous paragraph the relevance of the comparative methodological approach in social sciences and in particular in water studies, in the last decades, has been outlined. The methodological approach selected for the present research can be included in this comparative framework. Since the core of this research is the main framework of the water management reforms' paths and related issues in the Central Asian region after the collapse of the Soviet Union, two case studies were selected. This choice is based on the water dynamics that affected the region during the 1990s, as widely discussed in the previous chapter; therefore two different countries, Uzbekistan and Kazakhstan, were selected in order to make a comparative analysis of their water reforms' paths and related ways to the IWRM. Considering the framework and the characteristics of the research, a qualitative small-N approach was chosen; according to Ragin's position, as previously described, only a qualitative small-N methodology allows the understanding of particular sociopolitical and environmental processes as water management issues, and facilitates an in-depth analysis. Furthermore in the Central Asian context, where statistical data is not easily available and also is questionable, due to the historical/political environment of those republics, a qualitative small-N method offers the most effective and in-depth research approach for understanding the current ongoing processes. In contrast to comparisons of many countries through quantitative analysis, studies with a small number of cases are associated with fieldwork and a qualitative method of data analysis to gain a deeper understanding of processes instead of linking variables.[141] Moreover, being these water processes significantly affected by social issues, it seems difficult to obtain clear and complete results through a quantitative approach. Moreover, although a small-N comparison of two countries does not possess explanatory features for general theories, a structured in-depth comparison allows for bounded generalisations that might be tested on other cases or regions. This study employs one of the most classical methods in comparative analysis: a focused comparison using the Most Similar Case Design (MSCD); through this approach, dealing with differences in similar cases, the two case studies are selected on the basis of shared historical, social, political, and environmental features, so that the differences might be pointed out and highlighted. Before de-

---

[141] SEHRING, J., 2007. "cit.".

56

scribing in depth the steps of the field research, it should be clarified why these two countries, Uzbekistan and Kazakhstan, have been chosen for the research. The countries' importance (Uzbekistan and specifically South-Kazakhstan) in Central Asia for water use, irrigation, and in particular for the role of water in their political-economic systems have been the first criteria for the selection. These two countries are mostly located in the central-downstream section of the Aral Sea basin and they are crossed by the main rivers of the region, Amu-Darja (Uzbekistan), Syr-Darja (Uzbekistan and Kazakhstan) and Zeravshan (Uzbekistan), (FIG.3).

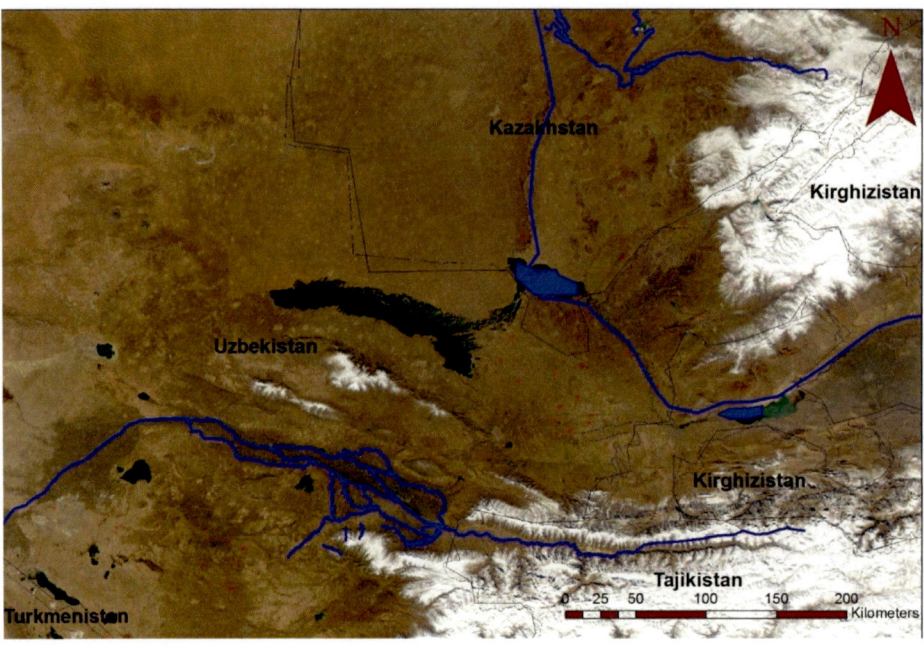

*FIG. 3. GIS elaboration of a satellite image (NASA-Modis,2003) representing the Syr-Darja river, crossing Uzbekistan and Kazakhstan, the Arys Valley (tributary of the Syr-Darja) and the Zeravshan valley and its canals. Blue lines: rivers and canals. Blue shapes: dams.*

Due to their agricultural emphasis on cotton, wheat, and rice farming, Uzbekistan and Kazakhstan use on average, as previously described, 88% of Syr-Darja's water amount and Uzbekistan uses a high rate (shared with Turkmenistan) of the Amu-Darja's one; the total irrigated area of the two countries reaches 5.6 million hectares (4.3 in Uzbekistan and 1.3 in South Kazakhstan, respectively), more than 65% of the total irrigated land in the Aral Sea basin.[142] Therefore, since Soviet times, these two countries, in particular Uzbekistan, have maintained a prominent position in international institutions and agreements concerning water — such as Interstate Commission for Water Coordination (ICWC) and Basin Water Organizations (BWOs) — and also in the water quotas disputes that occurred at the end of the 1990s. According to the Most Similar Case Design, it might be considered that Uzbekistan and Kazakhstan lie in the same geographical area. Furthermore, as two countries in the same area, they share a similar historical and cultural background — both were included in the Tsarist Russian Empire (since 1850–1860) — with specific features of water management practices and political legacies; also in previous times the political and religious framework did not present relevant differences among these two regions. In the last century the areas shared a common development under the Soviet Union, which transformed the society through the introduction of new institutions, policies, and practices, and shaping the attitude and the behaviour of the actors. Except for a few water facilities built in the last decade, the water infrastructures and the irrigation networks, inherited from the Soviet Union, do not present relevant technical and operational differences..Both in Uzbekistan and Kazakhstan, the first water laws issued at the beginning of the 1990s were mostly structured on the Soviet legacy. The different transition paths undertaken in the last decade, when the IWRM framework and the IMT process have been sponsored by international donors, will be analysed in the next paragraphs as part of the research core. Within the two countries, two valleys and connected irrigation schemes were selected for the field research: Middle Zeravshan valley in Uzbekistan and Arys valley in Kazakhstan. According to the administrative framework, the two regions refer respectively to the Samarkand province and to the South-Kazakhstan province. The two selected regions (Middle Zeravshan valley and Arys valley) do not present significant differences from a physical/geographical perspective, such as soil, water amounts, proximity to the mountains, annual precipitation and crops; these features hence are well suited for the MSCD approach. The only selection criteria was that the two valleys should lie in a representative area for Central Asia, hence a region characterized both by foothill areas and irrigation schemes. Although those characteristics are also present in other republics — for instance, in Kyrgyz and Tajik small plains — the foothill areas of Uzbekistan and Kazakhstan were selected by the Soviet

---

[142] IGC, 2002."cit.".

government as regions where design irrigation schemes oriented to supply cotton farming could be implemented. With the aim of providing an in-depth comparative analysis of the two water systems and to highlight the differences at the basin and local level, three districts within the provinces were selected:

- Urgut, Nurabad, and Pastdargom (Samarkand province) in Middle Zeravshan valley
- Tyulkibas, Ordabasy, and Otrar (South-Kazakhstan province) in Arys valley

These areas were selected by consideration of their physical position in the valley, upstream and downstream sections, and the proximity to the irrigation schemes. The basin/district level analysis in water studies has been widespread in the last decades, in particular focusing on developing countries. As claimed by Mollinga (2012), despite the fact that the main measures are decided at the national level, implementation is strongly related with the basins and its inner sociopolitical dynamics.[143] Also Wegerich (2012), focusing on the transboundary water issues, stated that the organizational and political dynamics occurring at the meso level can significantly differ from the national one, being strictly connected with the territories and the local environment.[144] The local level is the level of the lower administration that is responsible for the implementation of policy decisions carried out at national one; it is the immediate institution between the central political power and the target group — for instance, the water users. The social actors at the district/local level — the so-called "street bureaucrats" or "administrators-as-implementers"— are considered as important for the policy process as the top level: the institutional set-up responsible for implementation might thus be equally relevant for successful reforms as the central political powers.[145] In addition, among the different levels of an administrative system the district/local level is of special relevance in ensuring the implementation of formal regulations because it is at this level that formal and informal institutions meet.

---

[143] MOLLINGA, P. GONDHALEKAR, D., 2012. "cit.".

[144] WEGERICH, K. et Al., 2012. Meso-Level Cooperation on Transboundary Tributaries and Infrastructures in the Fergana Valley, *International Journal of Water Resources Development*, 28:3.

[145] SEHRING, J. , 2007. "cit." - WIMMER, A., DE SOYSA, I., WAGNER, C., 2003. *Political Science Tools for Assessing Feasibility of Reforms* (ZEF Discussion Papers on Development Policies, n. 63) Bonn.

Therefore, the implementation of national water policies and the international donors' supported measures at this scale among the selected districts will be outlined.

## 2.3 THE FIELD-RESEARCH STRUCTURE

Focusing on the field research, the study combines deduction of hypotheses from the conceptual framework with an inductive approach through the interpretation of the empirical data. The field-research period was therefore split in four stays of several weeks (1 to 2 months) with periods ranging from three to six months, which were used for the analysis of the first collected data and for the deepening of the research perspective. Such a cyclical field-desk process allowed a deep reflection on the data and a deep analysis of the main research carried out in the field of basin/district level water studies. The total field-research period was seven months: one month in spring 2011 (Uzbekistan), two months in autumn 2011 (Kazakhstan), two months in spring 2012 (Kazakhstan and Uzbekistan) and two months in autumn-winter 2012 (Uzbekistan and Kazakhstan). As claimed by Mollinga (2008), water crises are often not strictly related with water scarcity, but with lack of management.[146] Therefore, the focus of the research, being structured through a human/political geographical perspective is not based on the natural aspects of water, but on human actions, social and political. The research aims to gain an understanding of the social-political processes and dynamics connected with the IWRM and IMT implementation and its logics at the basin/local level in the current transitional context where governments and international donors are both engaged. Consequently, qualitative methods are the best methodological approach to examine these issues and the actors involved in these processes. As stated by Sehring (2007), qualitative methods are more applicable in the context of developing and transitional countries, such as those in Central Asia, than quantitative ones, because social issues are being dealt with and because it is challenging to get official data, which can be questionable due to the political and economic national contexts.[147]

---

[146] MOLLINGA, P. 2008. "cit.".
[147] SEHRING, J. 2007. "cit."

## FIELD - RESEARCH PERIOD (2011-2012)

| Research period | April '11 | Sept-Oct '11 | Apr-May '12 | Oct-Dec '12 |
|---|---|---|---|---|
| **Research places** | Uzbekistan<br><br>Tashkent-Samarkand Urgut-Nurabad distr. | Kazakhstan<br><br>Almaty-Shymkent Ordabasy distr. | Kazakhstan<br><br>Shymkent Tyulkibas/Ordabasy Otrar districts<br><br>**Uzbekistan**<br><br>Samarkand /Pastdargom | Uzbekistan<br><br>Tashkent Pastdargom-Urgut<br><br>**Kazakhstan**<br><br>Tyulkibas/ Ordabasy Otrar districts |
| **Research aims** | Experts interviews (national / district level)<br>Field - surveys interviews | Experts interviews (national / district level)<br>Field - surveys interviews | Field interviews | Experts interviews<br>Data analisys<br>Interviews |

Through a qualitative approach, the researcher has the possibility to choose which topics to explore in depth, depending on the actor or institution involved, the particular aim, and the social environment. As far as the specific qualitative approach, the following methods were used: semi-structured interviews, open interviews, informal conversations, and field surveys and walks. The semi-structured interviews were carried out both with "institutional" experts and with the water users, although with differences. The semi-structured interviews with the experts aim at exploring the context of the research, at delimiting its "boundaries", and generally at the understanding of the main issues and current policies related to the research questions; the people interviewed are not those with stakeholders' interests, such as water users, but as experts on a specific issue. An expert can be defined as an actor that is in charge of national or province/district level policies, or with decision-making processes; experts are able to provide the knowledge for the structuring of the research, being involved in the technical aspects or the political process or in charge of relevant social issues. For this research, experts were considered those who deal with water management and agricultural issues in government institutions both at the national and province levels, such as ministries, water and agricultural departments, and basin agencies as well as academic members working in universities and in the academies of sciences in the fields of water problems, development studies, geography, and history. Other experts, strictly dealing with social/development issues and natural resources management are members of international organizations, such as development agencies, often collaborating on environmental or sustainability projects. Interviews with these actors were designed as semi-structured, focusing on few and relevant topics and not defined questions, giving the interviewees the possibility to decide where to focus, and to talk and discuss in a open way. Concerning the language, interviews were carried out in different ways: most of them were in Russian or Uzbek, where an interpreter, often friends or students, helped in the translations, and others were conducted directly in English or Russian; interviews were conducted face-to-face, not via e-mails, Skype or telephone. The selection criteria for the experts to be interviewed were oriented at the first step towards the research questions and the understanding of the sociopolitical context; at the second step several interviews were carried out to clarify data previously collected or ambiguous statements regarding a specific topic. Furthermore, the snowball system was used in order to create a network in the institutional/academic environment and to compare knowledge and opinions on the specific research topics. In some situations the semi-structured interviews were followed by informal talks, often coinciding with changes in the physical or social environment; these informal discussions greatly facilitated in-depth understanding of some specific details or personal ideas and experiences. During the last field trip, six weeks were spent in the International Water Management Institute (IWMI) in Tashkent (UZB), collaborating with the IWRM's Fergana Valley project on the WUAs' changes. This period was extremely helpful in clarifying some issues at the end of thefield research. In the following table an overview of the experts/institutions interviewed is presented:

| Country | Uzbekistan | Kazakhstan |
|---|---|---|
| **Organization / Institution** | Academy of Sciences-Institute of History | Institute of Geography-Almaty |
| | Academy of Sciences Institute of Archeology | Al-Farabi Uni- Faculty of Geography |
| | Academy of Sciences Institute of Water Problems | Shymkent Uni- Faculty of Geography |
| | Samarkand University-Faculty of Geography | Scientific Research Institute of Water- Taraz |
| | Ministry of Agriculture and Water Resources | Al-Farabi Uni- Faculty of Agriculture |
| | Interstate Commission for Water Coord. -ICWC | *Kazgiprovodkhoz* |
| | Institute for Irrigation and Melioration - SANIIRI | Balkash-Alakol Basin Agency -BVO- |
| | International Water Management Institute - IWMI | Aral/Syr-Darja Basin Agency -BVO- |
| | Centre for Economic Research -CER- | South-Kaz. Water Department |
| | Swiss Development Cooperation - SDC | SouthKazakhstan Hydro-geological State Ent. |
| | German Society for Development -GIZ- | Kazakh-German University, Almaty |
| | Zeravshan Basin Agency | |

Regarding the basin/district level case studies, as it was previously briefly described, two valleys were selected, Middle Zeravshan valley in Uzbekistan and Arys valley in South Kazakhstan, and within those territories, three districts for each were chosen for in-depth fieldwork (FIG.4).

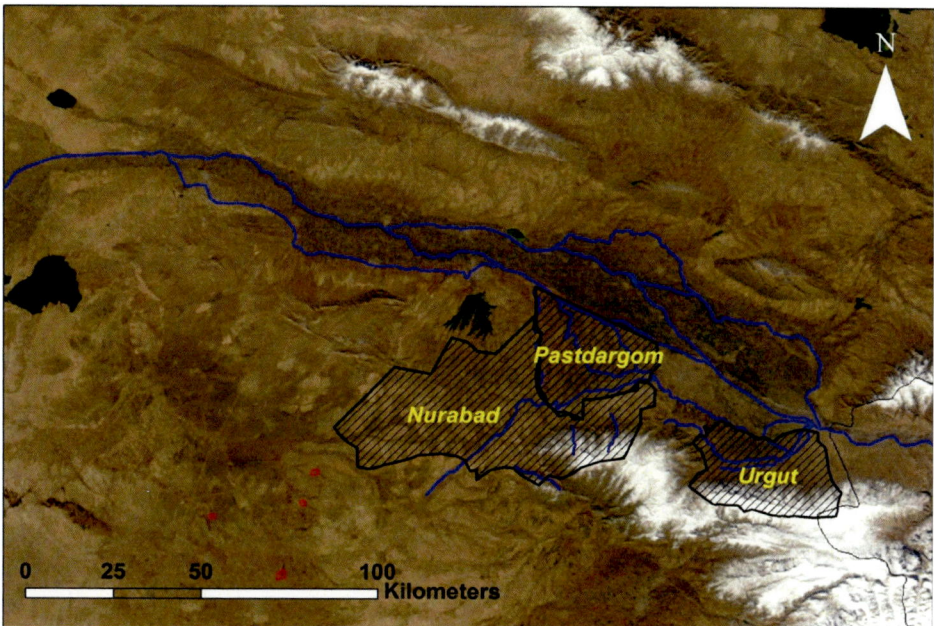

*FIG. 4: GIS elaboration of a satellite image (NASA-Modis,2003) representing the Middle Zeravshan Valley's irrigation scheme (flowing from the E to the W) and the three districts selected for the field research. Blue lines: river and canals. Black shapes: districts.*

The focus on this valley section, included in Samarkand province, was decided because it represents one of the largest and most important irrigated areas in Uzbekistan, and field research on water and agricultural issues was already conducted in this valley during my BA and MA studies. Therefore, knowledge of the territory's characteristics and personal contacts with several local experts and local institutions had already been established.[148] This condition was very helpful in creating strong relationships with several water users, facilitating and deepening the data collection through open interviews and informal talks. In the Middle Zeravshan valley, the districts (*raioni*) of Urgut, Pastdargom, and Nurabad were selected; those administrative units lie in the left-south bank of the Zeravshan River and are located along the river's valley upstream and downstream sections. The districts were chosen in order to have a complete representation of the valley — from the lands lying in the upstream part of the irrigation scheme

---

[148] Contacts with the Institute of Archeology of The Uzbek Academy of Sciences and with the Faculty of Geography (Samarkand State University) have been established since 2005.

to the central and the peripheral areas. In addition, in the Pastdargom district an international donor project involving WUAs and agricultural enterprises has been in progress since 2009, and therefore it was also selected to compare its water management's dynamics to those of the other districts. Moving to the other field-research case study, Arys Valley in Kazakhstan, this area was chosen because of its physical territorial similarities with the Middle Zeravshan valley, located approximately 400 kilometres to the south-east. These aspects will be more deeply described in subsequent paragraphs. In this valley too, three districts were selected in order to get an accurate comparison of the two field-research case studies using the same criteria: Tyulkibas district, lying in the upstream side of the valley; Ordabasy in the central part of the Arys' irrigation scheme (Arys-Turkestan canal); and Otrar in the downstream one (Shaulder' irrigation network). Focusing on the district-level research activity, in the Middle Zeravshan valley the study was formally supported by the Institute of History of the Uzbek Academy of Sciences in 2011 and by the International Water Management Institute (IWMI) in autumn/winter 2012; in Kazakhstan the fieldwork was sponsored (2011–2012) by the Faculty of Archeology, History and Geography of the Auezova State University based in Shymkent. The fieldwork had a starting point in each district, normally the district's main town, where the first contacts were made; consequently, the research was also carried out in other villages in order to get a complete overview of the local context. The districts' centres are the following: Urgut (Urgut district), Nurabad (Nurabad district), Juma (Pastdargom district)/Vanovka (Tyulkibas district), Temirlan (Ordabasy district), and Shaulder (Otrar district), (FIG.5). The methodological tools used during the fieldwork at the district level were semi-structured interviews, informal talks and field survey and observation. Just in some cases, as in Tortkol (Ordabasy district, Kazakhstan), the method of informal group discussion among cotton farmers was used. The fieldwork was structured both in Uzbekistan and Kazakhstan in different steps; semi-structured interviews with the main actors/experts at the basin/ provincial level followed the earlier step featured by data collection through interviews with experts at the national level.

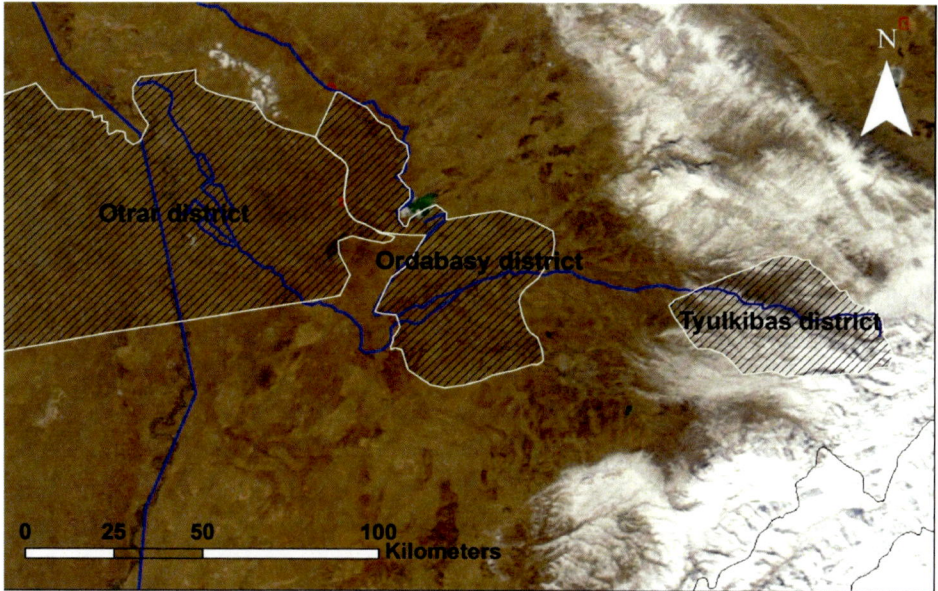

*FIG 5. GIS elaboration of a satellite image (source: NASA-Modis, 2003) representing the Arys Valley and its irrigation scheme and the three district selected as case-studies. Blue lines: river and canals. White shapes: districts.*

At the basin level semi-structured interviews were conducted with the Basin Agencies members in order to gather information about the basin's institutional and organizational structure (directors, administrators), as well as data regarding physical and technical features of the region and the water infrastructures, such as canals and reservoirs (hydro technicians). Afterwards, using the same methodological approach, the analysis shifted to the provincial water and agricultural departments *(Oblastvodkhoz/Oblastselkhoz)*, where the focus spanned data on water control and distribution, institutional and organizational differences within the Basin Agencies, land management, and agricultural issues. Both basins and provincial authorities were questioned about the transitional path of water reform, in order to understand their perceptions and then compare them with the data collected during the interviews with the experts. A great importance was given to the IWRM's pillars and related IMT issues and to the procedures for the establishment of the Water Users Associations, which has been the main issue of water management in Central Asia for a decade. Contrary to the interviews with the experts — parts of which were conducted personally in English or Russian — at the basin and local level, the fieldwork was completely carried out with the help of local interpreters, translating from Uzbek, Kazakh, or Russian to English or Italian. Subsequently, obtained the data on basin-province level, the focus was on the district water departments (*Rayvodkhoz* or *Kommunalnivodkhoz)* and on the Water Users Associations or WUAs (*AVP in Russian, SIU in* Uzbek, *SPKV* in Kazakh). Interviews were conducted with representatives of the WUAs, such as directors, administrators, hydro technicians, and accountants, in order to analyse the institutional, organizational, and territorial features, according to the IWRM pillars: that is, integration, participation of the water users in the WUA's decision-making as well as boundaries and technical issues. Also, attention was focused on the relations among the WUAs and the higher level authorities, basin agencies, and province departments. Although semi-structured interviews were conducted with WUAs members, in several cases, open interviews or talks allowed for more in-depth research. These different methods were possible when the social or environmental context changed — for example, during lunch or dinners with the WUAs members or during walks in the villages and along the canals. Those particular situations can often happen both in Kazakhstan and Uzbekistan when the researchers are able to be in contact at different times with the same persons. The same method — semi-structured interviews and informal talks — was also used for the final step of the fieldwork, carried out in villages randomly selected in the six districts. As a first step, interviews were conducted with farmers (*fermer),* owners of large plots and mostly oriented to cash cropping, and then to kitchen gardeners; the differences among these two types of farming will be analysed in the next chapters. With the villagers the focus was on water management issues at the village/farm level, as well as their perspectives on WUAs and the district water department's performance — for instance, the water supply practices, and their participation in the Water Users Association. Moreover some of them, depending on their personal availability, were encouraged to talk about about national water reforms, as well as their knowledge as WUA members, about the IWRM framework. As previously described, in some cases an informal group discussion with the farmers was conducted which provided a general perspective on the village's water managementdynamics. Furthermore, walks and surveys along the canals and in the fields were conducted with farmers and WUAs members, as hydro technicians or water managers, in order to see the canals' network and to understand the technical issues of water supply at the farm level. Regarding the empirical data analysis, part of it was conducted during the field-research months, in order to immediate reflect on the collected data, enabling a potential deepening of knowledge or explanation in the following days, as well as at the end of the fieldwork. Qualitative content analysis was adopted, identifying the main contents of the collected data with a step-by-step reduction and dissection of the textual material. All the semi-structured interviews were transcribed, however, in-depth notes were taken during open interviews, as it was difficult to record the people interviewed, in particular in Uzbekistan. Consequently, the data was structured and deeply analysed, dividing it into organizational, socio-political, and technical categories, according to the example of the socio-technical analysis, enabling the comparative approach.[149]

---

[149] ABDULLAEV, I., MOLLINGA, P. 2010. "cit.".

# 3. WATER POLICIES IN THE CENTRAL ASIAN REGION: FROM THE HYDRAULIC MISSION TO THE IWRM

## 3.1 THE GEOGRAPHICAL OVERVIEW

Central Asian region can be considered the core of the Eurasian continent; due to climatic and environmental features, water represents by far a key element in the physical structure of the territory. Stretching from the Caspian Sea in the west to the Chinese Tian-Shan range in the east and from southern Siberia in the north to Afghanistan' plateau in the south, its territory nowadays politically includes the Central Asian republics of Former Soviet Union: Kazakhstan in the north, Kirghizstan in the north-east, Tajikistan in the south-east, Uzbekistan almost in the centre and Turkmenistan in the south-west (FIG.6).

*FIG. 6: Political map of Central Asia (source: United Nations) representing the boundaries of the Central Asian republics.*

Central Asia is included in the arid and semi-arid regions; it is dominated by deserts and steppes for most of its main expanse and by high mountains ranges and irrigated plains. Therefore, it can be physically divided into two parts, the downstream western area and the upstream eastern area: the western area mostly includes the lowland Ust-yurt plateau, featured by the Karakum and the Kizilkum deserts and by the Kazakh steppes. This arid region is delimited in the west by the Caspian Sea and in the south by the Kopet-Dag range. The upstream eastern section includes the Tian-Shan and Alay ranges in the north, and the Pamir plateau in the south as well as a transitional foothill area between the mountains and the steppes. This area, featured by several glaciers ranging on average from 3800 to 7500 metres above sea level (Pik Kommunizma -7495 m. a.s.l.-, Tajikistan) serves as the *chateaux d'eau* of the Central Asian rivers. Due to the physical conditions and distance from the oceans, the region is strongly affected by a harsh continental climate, characterized by low annual precipitation, excluding the high mountains, ranging from 400/500 millimetres/year in the foothill zones to 50/100 mm in the western deserts. Except northern and eastern Kazakhstan, which lie in the Arctic Sea's hydrographical basin, the Central Asian region is included in the large endorheic basin of the Aral Sea, covering 1.8 millions mq, crossed by the Amu-Darja and the Syr-Darja and other rivers. The basin can be described as a large drainage system that terminates in the Aral Sea which lies in the Ust-yurt plateau on the border between Uzbekistan and Kazakhstan.The mountain systems form the flow generation zone for the Central Asian rivers even though only Syr-Darja and Amu-Darja flow into the Aral Sea. The Syr-Darja river is the largest river in terms of length, measuring, from the headwater in Tian-Shan Kirghiz range, 3019 km with a total catchment of 219.000 km3; average annual water flow measures 38.3 km3. The Amu-Darja is the largest river of Central Asia, measuring 2540 km, with an annual flow of 73 km3, but with a catchment of more than 300.000 km3. Both rivers can be considered transboundary ones: Amu Darja, originating in the Pamir mountains, forms the border in the upstream valley between Tajikistan and Afghanistan while downstream it crosses Turkmenistan and Uzbekistan. The most important river of its catchment is Zeravshan, flowing from the Tajik Alay mountains to the Uzbek plains. Syr Darja river crosses Kirghizstan, Uzbekistan, and Tajikistan in the midstream area through the Fergana valley, the largest and most important intramontane basin of Central Asia. Downstream in Kazakhstan, before the confluence in the Aral Sea, Syr Darja river receives the flow of the Arys river. Other relevant transboundary rivers flow from the Tian-Shan ranges to the plains and finally to the steppes, such as the Cù and the Talas rivers, both flowing from Kirghiz Alay range to the Kazakh plains. According to Suslov (1961), most of the Central Asian rivers are fed by the meltwater from snow and glaciers. This produces two close but distinct flood seasons: one in spring (April and May) due to snow thawing, and the other in summer (July and August) due to glacial melting, both of which are favorable periods for irrigation. Due to the gradual slope of the plains, several rivers do not have a real valley in their middle course and during high water the channels in these sections have a width of 500 to 1500 metres. Most of the settlements in Central Asia lie in the foothill areas, stretching in an irregular belt from the Kopet-dag in the south-west to the Zaliski Alatau in the north-east, and in the downstream river deltas, such as Amu-Darja, Syr-Darja, and Zeravshan irrigated plains. In fact, due to the aridity of the climate, low precipitation, and high evapotranspiration rates, rainfed agriculture is possibleonly in the upper foothill areas. Due to an average of below 350 to 400 mm/y of rain, agriculture is not possible without irrigation.[150]

## 3.2 TRANSFORMING NATURE THROUGH WATER MANAGEMENT: THE HYDRAULIC TERRITORIES

Since ancient times in Central Asia, irrigation techniques evolved as a strategy for coping with the natural environment, and similar spatial patterns of irrigation development occurred at different times in areas with similar natural conditions.[151] River flows were diverted and several water infrastructures and irrigation canals were built in order to increase irrigated land. In the foothill areas there was a progression from naturally irrigated basins to artificial ones fed by small hydraulic systems; along the large lowland rivers, the earliest irrigation occurred in deltas in natural basins along the lower course of small tributaries,

---

[150] SUSLOV, S.P. 1961. *Physical Geography of Asiatic Russia,* WH Freeman, San Francisco.

SPOOR, M., KRUTOV, A. The Power of Water in a divided Central Asia.

BENJAMINOVIC, R., TERZINSKIY, P. 1975. *Irrigatzia Uzbekistana,* Nedatelstvo Fan, Uzbek CCP, Tashkent.

FEDCHENKO, F. 1870. "Topographical Sketch of the Zarafshan Valley, *Journal of the Royal Geographical Society of London,* vol.40, pp. 448-461.

RICKMERS, R. 1913. *The Duab of Turkestan: A Physiographic skecth and account of some travels,* Cambridge University Press, Cambridge.

LEWIS, R. 1992. *The Geographical Perspectives on Soviet Central Asia,* Routledge, London.

[151] ADRIANOV, B.V. 1995. The Influence of Economic Development in the Aral Region and its Influence on the Environment, *Geojournal,* 35.1, pp.11-16.

72

where later extensive irrigation schemes were created.[152] Natural environmental changes and the creation of extensive irrigation systems coincided with the rise of strong, centralized political powers in the region. According to Tolstov (1948), this political context was the key component for carrying out these major changes to river areas and territorial physical features that emerged during the Bactrian and Sogd empires.[153] For centuries Central Asia has been featured by traditions of hierarchy and authoritarianism among its settled populations along the river banks and in the oasis; authority is personalised and personal loyalties are deeply rooted—these characteristics fit the nature of an irrigated oasis society.[154] Khorezm/Karakalpakstan oasis, Zeravshan valley, Fergana valley, and the Kopet-dag foothills were the areas most affected by these territorial changes.

*FIG. 7: GIS elaboration of a satellite image (NASA-Modis,1999 – scale approx. 1:25.000.000, N↑) representing the main irrigated areas (yellow shapes) of the Central Asian region.*

[152] LEWIS, R. 1966. Early Irrigation in West Turkestan, *Annals of the Association of the American Geographers,* vol. 56, n.3.

[153] TOLSTOV, S.P. 1948. "cit.".

[154] AMINOVA, M., ABDULLAEV, I. 2009. "cit.".

Therefore, Central Asian territory is the result of the huge hydraulic projects and irrigation network construction carried out during the centuries, and can be considered as a patchwork of extended irrigated plains and fertile foothills surrounded by mountains, steppes, and deserts (FIG.7). In the modern period, most of the hydraulic projects that aimed to develop irrigated agriculture were initiated in the Tsarist Empire at the end of the nineteenth century. In the 1860s, when the Russian army conquered Central Asia and settled in the region, most of the old irrigation schemes built in ancient and medieval times had not been restored since the destruction of Gengiz Khan, while others were in operation and working successfully considering the tools available at that time.[155] In the following decades, after the decision of the central Tsarist government to develop cotton farming in Central Asia due to the favourable climate conditions, new hydraulic infrastructures were created and old irrigated areas were extended.[156] The main Russian projects were set up along the Syr Darja river, due to its significant and regular flow, specifically in the Fergana valley and in the Hungry Steppe. In the Fergana valley, where thousands of colonists settled, the irrigation canals built during the Khanat period were restored and cotton farmlands were extended. In contrast, the Hungry Steppe project was set up *ex-novo* in the plain lying on the west bank of the Syr-Darja river. Several canals were initiated in the second half of the nineteenth century but some ot them had to be abandoned due to lack of funds and technical issues. Despite ambitious plans, by 1916 only two canals were built (by Romanovskiy and Imperator Nikolaiy I) and a mere 35.000 ha were irrigated.[157] The trend towards increasing specialization in irrigated agriculture and in particular in cotton farming, initiated by the Russian empire, had been continued and strongly intensified during the Soviet Union. Although the total irrigated area in Central Asia sharply decreased due to abandon and damages during the October Revolution, by 1928 most of the lands were restored to irrigation, with progress increasing gradually during the 1930s. After the collectivization processes and the establishment of the collective and state farms — *kolchoz* and *sovchoz* — which occurred in the 1930s, total irrigated land in Soviet Central Asia and Kazakhstan reached 3.2 million ha.[158]

---

[155] MATLEY, I,. 1970. The Golodnaya Steppe: a Russian Irrigation Venture in Central Asia, *Geographical Review,* vol. 60, n.3.

[156] LEWIS, R,. 1992. "cit.".

[157] BICHSEL, C. 2012. "The Draught does not Cause Fear"- Irrigation History in Central Asia through James C. Scott's lenses, *Revuè des études comparative Est-Ouest,* vol.43, n. 1-2.

[158] LEWIS, R., 1962. The Irrigation Potential of Soviet Central Asia, *Annals of Association of*

Moreover, cotton farming became more and more important, changing the existing cropping patterns.

## 3.3 THE SOVIET HYDRAULIC MISSION

In the 1950s the Soviet centralized government began a significant hydraulic reorganization of the territory, focusing in particular on irrigated areas, which extended into the entire Aral Sea basin with the main aim of expanding cotton's monoculture. The main areas affected by these territorial changes through water control lay in Uzbek, Turkmen, and Kazakh SSR.[159] In the Uzbek SSR in the Fergana valley and the Hungry Steppe (Syr-Darja basin) and the lower Amu-Darja, and part of the Zeravshan valley (Amu-Darja basin) new irrigation schemes were designed, becoming the main irrigated areas for cotton farming. In the two areas already affected by the Tsarist empire's projects, new canals were built and the irrigated land greatly increased. In the Hungry Steppe, the Farhad reservoir on Syr-Darja was created in order to facilitate the diversion of the river's flow; new canals and pumping stations were built in order to also irrigate the foothills' higher areas, resulting in the total irrigated area increasing from 153.000 ha in 1948 to 350.000 in 1967, including new lands in South Kazakh SSR also.[160] In the Fergana valley, the South Fergana canal, the Aravan-Akbura canal and the Khodjabakirgan canal were built in the 1960s, extending the irrigated area from the centre of the valley, included in the Uzbek SSR, to the southern mountain slopes in Kirghiz and Tajik SSR.[161] In the lower Amu-Darja basin, the ancient oasis of Khorezm was significantly extended with the construction of the Tuyamuyun reservoir, diverting water from the river to a new irrigation scheme included in Uzbek and Turkmen SSR; the commanded area in the 1980s reached a total of 275.000 ha.[162] The Amu-Darja river underwent the most relevant changes in water diversion in the Aral Sea basin due to the construction of huge hydraulic infrastructures: in the southern bank the Karakum canal was created — spanning 1175 km- it is the longest irrigation canal in the

---

*American Geographers,* vol. 52, n.1.

[159] SPOOR, M., KRUTOV, A., 2003. "cit.".

[160] MATLEY, I. 1970. "cit.".

[161] DUKHOVNY, V. et Al. 2008. IWRM implementation: Experience with Water Sector reforms in Central Asia, in *Central Asian Waters (Rahaman, Varis),* Water and development publications, Helsinki University of Technology.

[162] HORNIDGE, A,K., et Al., 2011. Reconceptualizing Water Management in Khorezm, *Natural Resources Forum.*

world, arising from the middle Amu-Darja valley to the desert and providing water to the Murgab oasis and then along the Kopet-dag foothills.[163] This infrastructure was the key factor for the development of cotton farming in the Turkmen SSR.[164] In the northern bank, starting in the 1960s, huge pumping systems were created with the aim of irrigating the foothill areas and lifting water to the higher sides of the valley. A lifting system (200 km in length, 170 m in height) was built to increase the irrigated area of Kashkadarja and Surkhandarja oasis in southern Uzbek SSR. Downstream, the Amu-Bukhara lifting canal was created (100 km length, 50 m in height), transferring water from Amu-Darja to Bukhara province to support irrigation and cotton farming in the lower Zeravshan valley.[165] In total, in the low Amu-Darja basin the irrigated area doubled between 1970 and 1986.[166] Along the Syr-Darja river, in southern Kazakh SSR, the Chardara reservoir was built in order to regulate the river flow in the downstream valley. This region was the northernmost Central Asian area for cotton farming. Therefore, in the Arys valley the Arys-Turkestan canal and the Shaulder irrigation scheme were designed to transfer water from the river, irrigating 75.000 ha of new lands. Other commanded areas, oriented to grain farming, were created in Talas and Cu transboundary valleys, between Kazakh and Kirghiz SSRs.[167] Reflecting the strong Soviet power through territorial transformations and the extension of agricultural areas, at the same time in north-eastern Kazakh SSR, Krushchev promoted the Virgin Land Plan with the intent to increase the agricultural production of new land, a large steppe area (88 million ha), oriented to grain and wheat farming.[168] From a sociopolitical and economic perspective, these projects carried out by the Soviet central government in the 1950s aimed to legitimize the state's control of territory and water resources. In the Aral Sea basin, the USSR had to achieve one of the hugest cotton productions in the world, and so the Soviet government undertook these measures, demonstrating its power to rural livelihoods through water management and territory transformations.[169] In Soviet Central Asia and Kazakh SSR, the total irri-

---

[163] O'HARA, S., HANNAN, T., 1999. Irrigation and Water Management in Turkmenistan: Past Systems, Present Problems and Future Scenarios, *Europe-Asia studies,* vol. 51, n.1.

[164] BENJAMINOVIC R., TERZINSKIY L. 1975. "cit.".

[165] WEGERICH, K. 2011. Water Resources in Central Asia: regional stability or patchy make-up?,*Central Asian Survey,* 30:2.

[166] LEWIS, R. 1992. "cit.".

[167] WEGERICH, K. 2011. "cit.".

[168] LEWIS, R. 1992. "cit.".

[169] WEGERICH, K., 2006. *"Handing over the sunset"- External factors influencing the establishment*

gated land increased from 4.5 million ha in the 1960s to 8.8 millions of hectares in the 1990s; 4.5 million ha were included in Uzbek SSR, the republic most intensively involved in cotton farming.[170] According to Allan (2003), since the 1950s, the Soviet Union have started carrying out its hydraulic missionin the Aral Sea Basin, following the main water paradigm of that time. Though in most Western countries this process started changing in the 1970s, in the USSR it was carried out until the 1990s.[171] Water management and supply were seen solely in technical terms carried out according to a centralized approach, without considering any economic, social, and environmental issues; huge infrastructure projects like dams, reservoirs, and large irrigation schemes, relying on the belief in the technical possibility of completely controlling nature, were the hallmarks of this approach.[172]

## 3.4 THE WATER LEGACY IN THE USSR

Focusing on the Soviet water legacy, in the official water governance structure of the Soviet Union all waters were centrally managed in Moscow by the Ministry for Land Reclamation and Water Resources (*Minvodkhoz*).In Central Asia, a regional agency (*Sredazminvodkhoz*) was responsible for the irrigation water of the whole Aral Sea basin, under the control of the central authority. This is often mentioned as a positive aspect as it led to a basin-wide approach with integrated water and energy management. In fact, most of the irrigated areas and related water use lay in downstream riparian republics (Uzbek, Turkmen, and Kazakh SSRs), while the upstream ones (Tajik and Kirghiz SSRs) were much involved in hydropower production. The Central Asian republican institutions and interests in resource utilization were subordinated to the central authority in Moscow, the Russian and Soviet capital, and to the greater interest of the USSR.[173] According to Renger (1998) the Ministries of the Central Asian republics were extensions of the Ministry in Moscow; they were responsible for fulfilling the centralized plans and norms and their role in decision-making was limited to providing data to the centre". The subordination of the republics was two-fold: sec-

*of Water Users Associations in Uzbekistan: Evidences from Khorezm Province,* Phd thesis, Humboldt Universitet zu Berlin.

[170] SPOOR, M., KRUTOV A., 2003. "cit.".

[171] ALLAN, T., 2003."cit.".

[172] SEHRING, J.,2007. "cit.".

[173] WEGERICH, K.,2006. "cit.".

toral (with regard to irrigated agriculture) and national.Therefore, the republican Ministries of Water Resources were merely implementing the decisions made in Moscow. Consequently, the utilization of the rivers did not correspond to the interests of the administrative zones.[174] Despite the Central Asia regional agency, based in the Aral Sea basin, water management and allocation was managed according to administrative boundaries, standardized with fixed schedules (quotas) for the republics, provinces (*oblast*), and districts *(rayon)*; hence, the provinces' water departments *(Oblastvodkhoz)* and district water departments *(Rayonvodkhoz)* were widely created in the whole Soviet Union. The administration of Soviet water management itself was featured by a fragmentation of many subordinated agencies to *Minvodkhoz* that lacked clear allotment of competencies; overlapping functions led to inconsistencies and weak implementations. Moreover, the USSR *MinVodKhoz* combined the planning, supplying, receiving, and controlling functions in one agency with minimal external control. Consequences were, on the one hand, a fixation on reclamation and construction projects rather than on operation and maintenance (O&M), and on the other hand, poor quality work in order to meet production plans.[175] At the local level, when collective farms were created, communal water management was centralized; the administration of the *kolkhoz* and *sovkhoz* was responsible for O&M of the on-farm systems. Theoretically, the collective members' meeting decided on water allocation among the agricultural brigades, but informally the measures were taken according to the production targets by the collective management together with the district water department *(Rayvodkhoz)*.[176] The Soviet ideology of total human command of nature led to a belief in unbounded exploitation of water resources; all water bodies were considered to be under state ownership; therefore at the farm level, the water users did not have to pay for water use. These irrigation practices, claimed by the Soviet government, led to huge water consumption without considering any economic or environmental issues. However, pre-Soviet times should not be idealized: rent seeking in water allocation was a tradition in Central Asia that predated the Soviet Union by centuries.[177] According to Wegerich (2006), even though these [Soviet] policies either indirectly influenced or directly altered the institutions responsible for local water manage-

---

[174] WEGERICH, K.,2006. "cit.".

[175] THURMAN M. 2002: Irrigation and Poverty in Central Asia. A Field Assessment. http://www-esd.worldbank.org/bnwpp/documents/7/IrrigandPovertyInCAvers2.pdf.

[176] SEHRING, J. 2007. "cit.".

[177] SEHRING, J., 2007. "cit.".

78

ment, these institutions kept to a certain pattern; this pattern was the system of patronage, widespread in Central Asia since the ancient oasis societies.[178] In the 1980s significant disputes emerged among the Central Asian republics regarding the different and unequal water use of Amu-Darja and Syr-Darja rivers. As analysed above, within the Aral Sea basin framework, dams and reservoirs were built upstream in the mountains of Tajik and Kirghiz SSRs, while the irrigation areas were downstream in the valleys and in the plains. The water management infrastructures were built to enhance irrigation in the downstream regions, with the aim of regulating the flows, releasing water during the vegetation period and retaining it inwintertime.[179] Therefore Tajik and Kirghiz SSRs had the possibility to use a small amount of water, originating in their territories, compared to Uzbek, Turkmen, and Kazakh republics, where most of the cotton production of the Soviet Union was carried out. Along the Syr-Darja river these issues clearly emerged after the construction of the Toktogul reservoir in Kirghiz SSR territory in 1974, which led to a deterioration of relations between the Kirghiz republic on one side and Kazak and Uzbek on the other.[180] In order to face the growing mistrust over water allocation and management between the Central Asian republics, Basin Water Organizations (BVOs) were created. The Syr-Darja catchment was totally included within the Soviet Union's territory; therefore, it was possible to manage the river and to establish a new organization according to hydrological boundaries. In 1984, Protocol n. 413 of the Ministry of Land Reclamation and Water Resources provided water distribution limits for the Syr-Darja.[181] A few years later, in 1987, the water limits were set for the Amu-Darja and through Protocol n. 566, BVO Amu-Darja was established.

---

[178] WEGERICH, K. 2005. *Institutional changes in Water Management at local and Provincial level in Uzbekistan,* Bern et. Al.

[179] RAHAMAN, R. 2012. Water Wars in 21st century along International River Basins: speculation or reality? *Int. J. Sustainable Society,* vol. 4, n.1/2.

[180] JOZAN, R. 2008. "Etat délinquant" ou modèle deviant? Retour sur le non respect de traite international de partage de la ressource en eau du Syr-Darja, *Metropolis/Flux,* 1, n. 71.

[181] DUKHOVNY, V., SOKOLOVV. 2005. "cit.".

-------------------------------------------------------------------------------

**SOURCES OF RIVER FLOWS IN THE ARAL SEA BASIN (annual averages in km 3)**

| Country | River Basin | |
|---|---|---|
| | Syr Darya | Amu Darya |
| Kazakhstan | 2.4 | - |
| Kyrgyz Republic | 27.6 | 1.6 |
| Tajikistan | 1.0 | 49.6 |
| Turkmenistan | - | 1.5 |
| Uzbekistan | 6.2 | 5.1 |
| Afghanistan and Iran | – | 21.6* |
| Total for Aral Sea Basin | 37.2 | 79.3 |

Source: Wegerich K., 2006

-------------------------------------------------------------------------------

Contrary to the Syr-Darja BVO, this basin could not be based on hydrological boundaries because Afghanistan, where part of the flow originates, was not involved in the organization. Despite the aim of the protocol, the water quota for Uzbek and Turkmen SSRs reached 83% of the total amount.[182] The limits for the different republics did not reflect equitable water resources use, but highlighted the subordination of all the republican institutions for water resources utilization to the central government in Moscow and to the greater interests of the Soviet Union. Irrigated agriculture used on average 90% of the total Aral Sea basin's annual water resources; in the Amu-Darja basin, the irrigated area was divided as follows: 22.000 ha in Kirghiz SSR, 469.000 ha in Tajik SSR, 2.300.000 in Uzbek SSR and 1.700.000 in Turkmen SSR.[183] At the end of the 1980s, with the aim of reducing these inequities, a unified water-energy system was established among upstream and riparian republics in the operative framework of Syr-Darja BVO; downstream countries had to pay upstream ones for the summer release of water stored during wintertime with free gas and coal to generate electricity in

-------------------------------------------------------------------------------

[182] WEGERICH, K. 2006. "cit.".

[183] DUKHOVNY, V. , DE SCHUTTER, J., 2011. *Water in Central Asia: Past, Present and Future*, CRC press.

the cold winter months.[184] In the Amu-Darja BVO no measures were undertaken. Since the 1960s in the whole Aral Sea basin intensive irrigated agriculture
and unsustainable use of water resources — featured by over irrigation, dissipa-
tion in the supply networks, and dysfunctional drainage systems — caused several significant environmental problems. The significant reduction of Amu-Darja
and Syr-Darja annual flows into the Aral Sea, due to the diversion of the waters,
led to a progressive shrinking of the lake.[185]

**INFLOW INTO THE ARAL SEA** (km$^3$) Source: Lewis, R.1992

| Years | Amu-Darja | Syr-Darja | Total |
|-------|-----------|-----------|-------|
| 1960 | 37 | 21 | 58 |
| 1965 | 25 | 4 | 29 |
| 1970 | 28 | 9 | 38 |
| 1975 | 10 | 0.6 | 10.6 |
| 1985 | 7.9 | 0 | 7.9 |

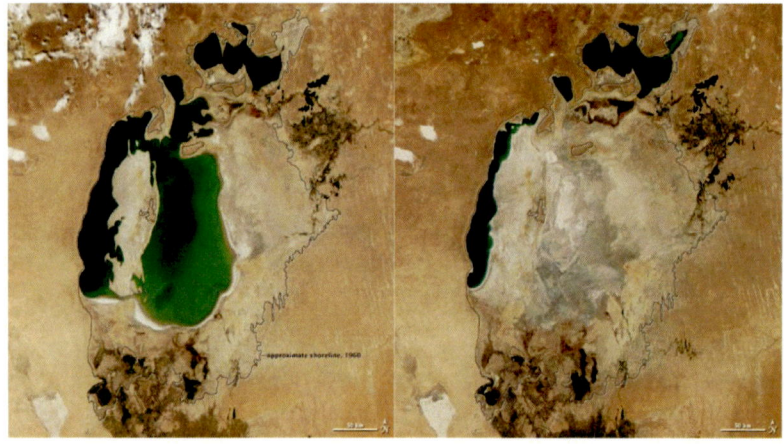

*FIG.8.: The Aral Sea shrinking: comparison of two satellite images (NASA-
Modis, 2014) of the sea (left: 2000, right: 2014) ( approx. scale 1: 8.000.000,
source: www.sciencedaily.com).*

[184] GRANIT, J. Et al. 2010. *Regional Water intelligence: report on Central Asia,* UNDP-SIWI, paper
n. 15.
[185] MICKLIN, P. 2007. The Aral Sea Disaster, *Annual Review of Earth and Planetary Sciences,* 35.

Featured by high salinity, the Aral Sea area has been nearly halved since the 1960s and its shores have retreated in some areas more than 120 km; since 1989 the northern part of the lake has separated from the larger southern part due to desiccation, creating two water bodies, the small and the big Aral Sea (FIG.9).[186] The Syr-Darja flows into the northern part, while Amu-Darja flows into the big body. A channel (river), flowing from the north to the south, connects the two lakes, particularly during the snow-melt season.[187] The areas affected most by the Aral Sea's desiccation were the neighbouring Amu-Darja delta in Kazakh SSR and the Karakalpakstan autonomous republic in Uzbek SSR, where economic activites, such as fishing and kitchen gardeners' agriculture, significantly deteriorated. In the decade from 1980–1990, this ecological issue was almost obscured by the Soviet central government and the Central Asian agencies in order to pursue cotton farming plans; it was only just before the USSR's collapse that these issues started to emerge and were discussed.[188] As analyzed by Lewis (1992), most of the authorities directly involved in water resource management and irrigation provided astonishing statements about the Aral Sea issue; the President of Turkmen SSR's Academy of Sciences claimed that the drying out of the Aral could be more advantageous than its preservation by supplying new fertile land; moreover, it was asserted that the disappearance of the sea would not lead to any negative environmental consequences in the region.[189] This statement was confirmed by the Prime Minister of the Ministry of Land Reclamation and Water Resources who added that "Aral should die beautifully", clearly justifying the means for obtaining the cotton plans and irrigation projects. Due to the evident environmental problems related to the Aral Sea's desiccation, a few years later, the Soviet authorities finally admitted the existence of a serious problem. Most of the irrigated areas were affected by high salinity due to a lack of crop rotation and archaic drainage networks that could not handle the problem of water logging.[190] Therefore, a lot of water was used to clean the soil at the beginning of the vegetation season. This problem particularly affected the downstream areas where the high salinity rate caused a significant reduction of the bio-diversity in the wet areas, such as the *Tugai* forests located along the riverbanks. Furthermore, the Aral Sea's shrinking significantly contributed to the

---

[186] LEWIS, R. 1992. "cit.".

[187] MICKLIN, P. 2007. "cit.".

[188] MAINGUET, M. 2007. *L' Homme et la Secheresse*, Masson, Paris.

[189] LEWIS, R., 1992. "cit".

[190] SPOOR, M., KRUTOV, A. 2003. "cit.".

climate change in the surrounding areas leading to an increase in the length of the frost season in winter and a rise of average temperatures in summer.

## 3.5 THE POST INDEPENDENCE REGIONAL WATER CHALLENGES

The collapse of the Soviet Union led to institutional changes and the creation of inter-republican agreements regarding the transboundary rivers issues, water use, and the Aral Sea. The shift from one administrative unit, the Soviet Union, into five independent countries (Kazakhstan, Kyrgyzstan, Uzbekistan, Tajikistan, and Turkmenistan) led to major challenges regarding transboundary water use and to the Aral Sea basin management in general.[191] At the beginning of the 1990s, the regional approach of water management inherited from the Soviet Union was at risk and the Central Asian region was considered, in the international agencies' scenario, a potential crisis area in terms of natural resources management. Nevertheless, soon after the independence achieved in 1991, the governments of the new Central Asian countries agreed to continue with the principles of water allocation that prevailed during the Soviet Union, hence continuing to support the established basin agencies, meaning BVO Syr-Darja and BVO Amu-Darja. In order to strengthen the international water resource framework, in 1992 the five Ministries of Water Resources of the new Central Asian independent countries met in Almaty (Kazakhstan) in order to sign an agreement ensuring cooperation in joint management, use, and protection of interstate sources of water resources; this agreement founded a joint authority named "Interstate Commission for Water Coordination (ICWC-MKVK).[192] Under the agreement, the state retained their Soviet-period water allocation, refrained from projects infringements on other states, and promised an open exchange of information.[193] In addition to the water quotas, the ICWC was established in order to regulate and ensure equal management of the main water infrastructures located on the Amu-Darja and Syr-Darja rivers — such as the Toktogul, Chardara, and Kairakkum reservoirs — aiming to prevent potential disputes. BVO Syr-Darja and Amu-Darja were included in ICWC as executive bodies in charge of planning and managing water flow regimes and distribution, as well as directly implementing the decisions made by ICWC relevant to water allocation, schedules

---

[191] DUKHOVNY V., SOKOLOV V.V. 2005. "cit.".

[192] DUKHOVNY, V. et Al. 2009. *Integrated Water Resources Management: Putting Good Theory into Real Practice. Central Asian Experience.* SIC ICWC, GWP CACENA, Tashkent.

[193] WEGERICH, K. 2006. "cit.".

of water flow and releases, and water quality control.[194] Furthermore, in the framework of ICWC, the Scientific Information Centre (SIC) was created to develop methods and approaches for prospective development, improvement of water management, and safeguarding the ecological situation in the basin, working in close collaboration with networks of scientific organizations. Although this international framework was established to improve interstate water resource management and allocation, the evidence showed that, on the one hand, it prevented severe disputes but, on the other hand, no real improvement had emerged, partly due to lack of coordination. Numerous unresolved tensions among Central Asian countries continued to rise.[195] These issues emerged in the small transboundary river's irrigated area, excluded from the main international agreements, and also in the Syr-Darja basin where the water shares have not changed since the establishment of BVO in 1984. Regarding water quotas, Dukhovny and Sokolov (2005) argued that there was no way to preserve the desired "status quo" of former water allocation and use because of the emerging political and economic differences in development strategies among the Central Asian countries. Moreover, the reduction of funding for operation and maintenance was common after the USSR's collapse, leading to a deterioration of water infrastructures and related problems in management and control.[196] Even though the newly independent countries decided in 1992 to share water resources according to the Soviet agreement, other regional institutions and practices, such as the exchange of food and energy, disappeared; hence each of the new countries had to develop their own challenging strategy of energy and food security. While the downstream countries could divert water away from cash production (cotton) to food production (wheat and rice), the small amount of water allocated to upstream countries, such as Kyrgyzstan, did not allow important changes.[197] This policy was in effect in the 1990s in Uzbekistan, where wheat farming significantly increased to cope with food self-sufficiency. The potential rise in water demand in the upstream countries for agricultural purposes inevitably led to a decrease inwater availability in downstream countries. Despite the deterioration of water infrastructures, after gaining independence, several countries — in particular Turkmenistan, Kazakhstan, and Tajikistan — extended their

---

[194] DUKHOVNY, V. et Al. 2009. "cit.".

[195] ALLOUCHE, J. 2007. The governance of Central Asian waters: national interests versus regional cooperation, (Central Asia at the crossroads), *Disarment Forum,* 4.

[196] DUKHOVNY, V., SOKOLOV V.V. 2005."cit.".

[197] WEGERICH, K. 2006. "cit.".

agricultural areas, generally increasing the total water use. Exceeding the water quotas was possible because of inadequate monitoring and lack of control; together, Uzbekistan and Kazakhstan used 88% (51% and 37%, respectively) of the Syr-Darya water flow, making agricultural development for upstream users difficult.[198] Furthermore, it was argued by the water community that the physical location of the ICWC and BWOs in Tashkent and Urgench (Uzbekistan) led to that country's advantages in water allocation, keeping the leadership role achieved during the Soviet Union through cotton farming.[199] The main tensions occurring in the mid-1990s between upstream and downstream countries was not based on an increase in upstream water demand but because of a shift in the dams' seasonal operations; the Syr-Darja BVO agreements' impact stopped functioning when Uzbekistan and Kazakhstan introduced market prices for petroleum and gas supply to Kyrgyzstan. Therefore, Kyrgyzstan started to release water from the Toktogul dam in wintertime to ensure energy-supply self-sufficiency, contrary to the interests of the riparian countries which relied on winter water storage and summer flow for irrigation.[200] In view of the weak and ineffective Aral Sea basin's governance system, according to Smith (1995), a new cooperation agreement based on the Syr-Darja's water releases should be signed in order to avoid further disputes and to keep regional stability.[201] On March 17, 1998, the Barter Agreement was signed between the governments of Kyrgyzstan, Uzbekistan, and Kazakhstan regarding use of the water and energy resources of the Syr-Darja river basin. According to the agreement, Kazakh and Uzbek governments had to buy Kyrgyz electricity in summer and then sell gas and petroleum to Kyrgyzstan in winter; thus the summer water flow for cotton and the water supply for wheat in downstream areas would be ensured.[202] In 1999 the Barter Agreement was subscribed by Tajikistan too, as it is also involved in irrigated agriculture using Syr-Darja's flow in the southern part of the Fergana Valley. Although the Syr-Darja basin water use should be strengthened by this institutional framework signed by all the republics involved, the agreement had limited success. According to Wegerich (2006), it was less beneficial

---

[198] ICG, 2002.*Central Asia: Water and Conflicts,* ICG Asia report, n.34.

[199] WEGERICH, K. 2006. "cit.".

[200] JOZAN, R. 2008. "cit.".

[201] SMITH, D. 1995. Environmental Security and Shared Water Resources in Post-Soviet Central Asia, *Post-Soviet geography,* 36, 6.

[202] ZIGANSHINA, D. (no date). International Water Law in Central Asia: Commitments, Compliance and Beyond, *The Journal of Water Law,* 20.

for Kyrgyzstan because the price for hydropower was less than that for gas and petroleum, and also the provision that downstream countries receive the water at the time when it is needed led to additional income for them.[203] Furthermore, according to the agreement, the operation and maintenance costs of water infrastructures must be covered by the country where the facilities are located. Hence Kyrgyzstan did not receive any additional funds from riparian states to maintain dams and canals, although the Kyrgyz authorities tried to introduce an operation and maintenance charge.Moreover, as stated by Jozan (2008), although the agreement should be viable for a pluri-annual period, the Uzbek government did not accept this point, and, thus it had to be signed every year. Uzbekistan did not respect the inter-republican agreement and the following winters refused to pay the granted energy supply to Kyrgyzstan.[204] Despite this infringement, the water supply to Uzbek's irrigated areas in the Fergana Valley and the Hungry Steppe did not cease. The evidence showed that despite the institutional arrangements signed between the Aral Sea basin's republics, generally the state's internal interests prevailed in respect on cooperation among them. Furthermore, the inter-republican governance proved to be weak and ineffective in regulating the different states' interests in water allocation. Also, the unequal level of political influence among the Central Asian republics,emerged, reflecting the power distribution of the Soviet Union period. Since the 1990s the Aral Sea's desiccation too was considered an inter-republican issue to be dealt with and solved through cooperation and institutional frameworks; in March 1993, in Kizylorda (Kazakhstan), the Interstate Council on the Aral Sea (ICAS) was established by the Central Asian governments. The ICWC was placed under the new policy organ which in 1997 was transformed into the International Fund for the Aral Sea (IFAS).[205] IFAS is under the authority of the deputy Prime Minister of the Central Asian countries, excluding Afghanistan, as already occurred in the establishment of ICWC in 1992. Its status was stipulated in 1999 in Ashgabat (Turkmenistan) based on these aims: develop and finance environmental scientific and practical projects and programmes oriented to an environmental improvement in the areas affected by the Aral Sea disaster as well as to solve socio-economical problems in the region. The main IFAS task, in order to face these challenges, is to administer the Aral Sea Basin Program (ASBP), a framework designed with

---

[203] WEGERICH, K. 2006. "cit.".
[204] JOZAN, R. 2008. "cit.".
[205] ZIGANSHINA, D. (no date). "cit.".

the financial and coordination aid of the World Bank.[206] On the basis of the strategy, acts were signed in order to start applying measures oriented to the Aral Sea region's environmental and social protection. As previously analysed, starting from the 1990s, the Aral Sea divided into two bodies — the small one in the north (located in Kazakhstan) and the big one in the south (in Uzbekistan); in the last decade the southern one shrunk more rapidly compared to the Kazakh one. Since the 2000s water infrastructures were created, in particular in the downstream Syr-Darja valley, in an effort to increase the water flow into the small northern Aral Sea. At the end of 2003, the Kazakh government announced a plan for the construction the Korakal dam, conserving the Syr-Darja's water flow.[207] In 2005 the work was completed and since then the northern Aral Sea's level increased, while the lake's salinity generally decreased. According to the Kazakh Ministry of Foreign Affairs, the northern Aral Sea's surface increased from 2.550 kmq in 2003 to 3.300 in 2008 and the sea's depth increased from 30 to 41 metres. The project results and the lake's increase will hopefully,in the near future, lead to a rehabilitation of the Aral Sea's climate, featured by less extreme temperatures compared to the neighbouring Kyzilkum desert.[208] Contrary to the northern lake, despite of the interstate cooperation and the World Bank's support of the ASBP program, the southern Aral Sea has been largely abandoned to its fatefor the last couple of decades. Although no special program directly focused on this waterbody, the evidence confirmed that the Uzbek government did not support any measures to deal with the Aral issues. Only excess water from the northern Aral was periodically diverted into the largely dried-up Uzbek one through a sluice in the Korakal Dam; but a lack of political will has not yet supported this practice. In addition the massive water use of the Amu-Darja's flow to irrigate the cotton fields in downstream Uzbekistan has not been reduced in the last years.[209] The evidence clearly showed that despite the inter-republican agreements signed one decade ago, national interests have prevailed both from the political and economic perspectives. In addition the water use by the Central Asian countries in the Aral Sea basin is strongly linked with national water policies and the development path pursued since the 1990s. Although the actions of

---

[206] GRANIT et Al., 2002. "cit.".

[207] MICKLIN, P. 2007. "cit.".

[208] UNDP, 2007.*Integrated Water Resource Management and Water Efficiency in the Republic of Kazakhstan, 2008-2025*, Progress Report, UNDP PROJECT #35289.

[209] ABDULLAEV, I. et Al. 2009. Agricultural Water Use and Trade in Uzbekistan: Situations and Potential Impacts of Market Liberalization, *Water Resource Development*, 25, 1.

the international actors and donors have been determined since the USSR collapse, different policies between the republics emerged, significantly influencing water resources management.

## 3.6 THE ROLE OF THE INTERNATIONAL ORGANIZATIONS IN PROMOTING THE IWRM

### 3.6.1 The different national ways to institutional -water- reforms

The collapse of the Soviet Union led to several challenges in the Central Asian region connected with water management from a sociopolitical and economic perspective. The Soviet state-centralized system ceased to provide all the services in the agricultural and water sector, and the inter-republican barter agreements of economic and food and energy assistance also ceased to operate. Therefore, the newly independent countries had to reform the institutional and organizational pattern for natural resources and land management in the main framework of post-socialist transition, decentralization, and privatization processes. Although, as previously analysed, at the inter-republican level, the Almaty agreement (1992) was signed, maintaining the Soviet's water allocation quotas, at the national level each country carried out a different transition path strictly related with political, social, and economic internal affairs that were not necessarily consistent with one another. Having irrigated agriculture as the main and the largest income base, water became a crucial resource for the stability of the Central Asian region. The evidence from the early 1990s showed that it was totally unsustainable from multiple perspectives — technical, organizational, and environmental — to continue managing water in a state-centralized way down to the farm level. Moreover, at this level the inherited water management structure could not be applied in the changing context of the dismantling of the large collective farms, *kolchoz* and *sovkhoz*. This process occurred in the Central Asian countries starting in 1993, following different approaches and strategies, and was completed in shorter or longer periods; hence the newly formed independent states considered the shift from collective to private farms to be the best strategy for creating more advanced and productive units. These reform processes led to more freedom for individual farmers to grow crops of their own choice and reduced the state's role in input provision, including water. Changes in farm organization had a direct relationship with water management issues, increasing disputes and competition for water at the local level.[210] Despite these agricultural transition processes, the transformation of the republics' institutions has been assessed as gradual and state-centric. In some of the Central Asian countries, as will be analysed in this section, although water management went through different changes and transformations, it still remains more or less state-owned or state-managed at all levels. Water management also shows different dynamics at different hierarchical levels.[211] In other republics, the state directed the transition processes through a clear and defined privatization path, supporting the market economy's principles. These choices of development and of natural resources management were strongly determined by the political and economic will of the state apparatus and by the influence of international agencies in the state's internal affairs.

### 3.6.2 Supporting decentralization: the IWRM and the IMT

Regarding these major issues, in order to balance the different and some ways contrasting reform paths, since the early 1990s several international agencies began promoting neo-liberal policies such as market deregulation, privatization, and governance support in water resources management institutions. The aid agencies and development banks most involved in development support in Central Asia are the World Bank (WB) the Asian Development Bank (ADB), the International Monetary Fund (IMF), the United Nation Development Program (UNDP), the Swiss Development Cooperation (SDC), the German-Technical Cooperation (GTZ), theUnited States Agency for International Development (USAID), and others.[212] According to Molle et al. (2009), the international donors' main aim in Central Asia has been a structural adjustment of the institutional context supporting neo-liberal policies leading to a roll-back of the state as the central power.[213] In addition, Mollinga claimed that since 1992 a globalization of water resources management was promoted in several arid developing countries, according to the water framework discussed by the international water community in the Western countries.[214] Merrey (1996) added, focusing on the transition processes, that governments should leave water management control of the on-farm irrigation schemes to the water users in order to increase respon-

---

[210] ABDULLAEV, I. 2009. "cit.".

[211] AMINOVA, M. ABDULLAEV, I. 2009. "cit.".

[212] GHAZUOANI et. Al, 2012."cit.".

[213] MOLLE, F. MOLLINGA, P. and WESTER P., 2009."cit.".

[214] MOLLINGA, P., GONDHALEKAR, D. 2012. "cit.".

sibility and sustainable water use.[215] In addition to the neo-liberal institutional and organizational approach, the action of the international agencies focused on sustainability. Through these approaches, the donors and the development banks encouraged the Central Asian institutions to accept and implement the concepts of participation in the decision-making processes, integration of water use, equitable and sustainable water allocation practices, and generally to promote the "economic value" of water by introducing water fees systems. These principles of water resources management constitute the backbone of the Integrated Water Resource Management (IWRM), the water paradigm supported worldwide by the international water community since the mid-1990s. According to the development agencies' strategy, the starting point for implementation of these principles supporting the IWRM framework should focus on managing water according to hydrological boundaries and equitable utilization of water resources. This presented a great challenge considering that — excluding the Syr-Darja BVO, which was based on hydrological boundaries — the whole Aral Sea basin water resources had been managed since the Soviet times according to administrative boundaries; in addition the principle of equitable utilization of water was quite far from the Soviet water management system. As previously analysed, during the Soviet Union era, water resources had never been viewed as an economic good, hence they were used without considering any environmental or sustainable approach. Therefore it was crucial for the donors to determine whether the sociopolitical and economical environment of the newly independent Central Asian countries could fulfil the requirements of undertaking the transition path. According to Dukhovny and Sokolov (2006), in the process of reform implementation, some relevant key points have to be considered, such as the political environment, economic choices, the water facilities' conditions and the social groups' needs.[216] Moreover in many societies, in particular where the governments had a strong role in water control, the political powers were seen as the ultimate provider of such services and there was reluctance on the part of the farmers to take over such responsibilities. Despite these issues, as occurred in several arid developing countries, the international donors — in particular, the WB, the ADB and the USAID — supported the establishment in Central Asia of water institutions based on hydrological boundaries (basin agencies). At the lo-

---

[215] MERREY, D. 1996. *Institutional Design principles for Accountability in Large Irrigation Systems,* International Irrigation Management Institute (IIMI), research paper 8.
[216] DUKHOVNY V, SOKOLOV VV. 2006. "cit.".

cal level, the transfer of water facilities from state responsibility to farmers organizations' control (Irrigation Management Transfer (IMT) — with the aim of improving the accountability of the irrigation service to farmers and making that service more productive and sustainable — was also supported.[217]

### 3.6.3 The widespread adoption of the WUAs at the local level

To institutionalize the IMT, the Water Users Associations (WUAs) establishment was strongly supported through development programs and financial aid. Concerning the different types of WUAs promoted and established worldwide, in the Central Asian context the WUAs generally refer to farmers' associations corresponding to the secondary or (more frequently) to the tertiary level of the irrigation systems, depending on the infrastructures and water management features.[218] Salman (1997) states that through the establishment of WUAs it is possible to achieve optimal utilization of available water through a participatory process that endows farmers with a major role in management decisions over water in their hydraulic unit.[219] Furthermore, the World Bank lists a number of benefits which participation of users in managing and maintaining water facilities may bring: benefits include (i) increasing the likelihood that these water facilities will be well maintained, (ii) contributing to community cohesion and empowerment in ways that can spread to other development activities and (iii) reducing the financial and management burdens on the government as a result of users' participation in management and maintenance of such water facilities. The WUA establishment process can be hampered not only by political resistances but by cultural ones; since ancient times water in Central Asia,has been considered a God-given resource for which no fees should be levied, and so in this region the establishment of WUAs could be seen by farmers as a process for facilitating the levying of such fees, or for ending other subsidies provided to them. Furthermore, this idea was culturally enforced during the Soviet Union, when all the costs for water allocation and use were covered by the state and its agencies. According to the donors' rationale in Central Asia, the supported WUAs' performance should be based on a strong legal framework and on three

---

[217] GHAZUOANI et Al. 2012. "cit.".

[218] GUNCHINMAA T. ,YAKUBOV, M. 2009. Institutions and Transition: does a better institutional environment make the water users' associations more effective in Central Asia?,*Water Policy*, 1, 22.

[219] SALMAN, M.A, 1997. *The Legal Framework for Water Users Associations- A Comparative Study,* World Bank technical papers, n.360.

main domains of responsibility: water management, facilities' maintenance, and financial costs management.[220] The legal framework consists basically of three sets of legal instruments: the enabling law, the bylaws of the WUA, and the transfer agreement between the irrigation agency and the WUA. In fact, focusing on the domains, the water management inside the WUAs should be scheduled according to a joint relation between the farmers involved in water allocation at the farm level and the irrigation agency, providing them with water quotas. In the field of facilities' maintenance, the WUAs' employees — in particular, hydro technicians and water-allocation workers — would be responsible for the sluices, the hydro posts and the small canals without any interference from governmental agencies. This domain would be strictly connected with financial costs management: all the expenses for technical and organizational matters should be covered by the WUAs' members through the irrigation service fee (ISF).[221] Yakubov (2012) argued that, due to multiple donors' support for Irrigation Management Transfer (IMT) and the WUAs' establishment, and allowing for national adaptations, quite a confusing mix of different WUAs models have emerged. These organizational structures and practices have been set up in most countries of the region without anyone really knowing what has worked and what has not — such as top-down technocratic WUAs versus bottom-up and user-participative ones; WUAs organized along hydrological boundaries versus those following the administrative boundaries of the former collective farms or even former administrative districts. All of these established associations have also varied in terms of size, scale, tariff rates, funding, project duration, and modes of engagement with other WUAs and the basin/regional government authorities.[222] Yalcin and Mollinga (2007), in their review of the WUAs-related project in Uzbekistan, stated that more than ten projects funded by seven different donors were set up in the last decade; the same dynamics have also occurred in the other republics.[223] Therefore the institutional reforms in the Central Asian region differed greatly in terms of trajectories, features, and times, and the

---

[220] SALMAN, M.A, 1997. "cit.".

[221] GHAZOUNANI et Al., 2012. "cit.".

[222] YAKUBOV, M., 2012. A Program Theory Approach in Measuring Impacts of Irrigation Management Transfer Interventions: The Case of Central Asia, *International Journal of Water Resource Development,*28:3.

[223] YALCIN, R., MOLLINGA, P. 2007. *Water Users Associations in Uzbekistan.The Introduction of a New Institutional Arrangement in Local Water Management.Amu-Darja case study, Uzbekistan,* NeWater project, University of Bonn.

WUAs' establishment had a different impact on the various republics' water management being carried out in strict relation with the land and agricultural' reforms. [224]

## 3.7 THE INSTITUTIONAL WATER REFORMS IN THE CENTRAL ASIAN STATES

Focusing on the regional context, the Kyrgyz republic has been at the forefront of the institutional reforms in the Central Asian region both in the agricultural and water sector. After independence, the Kyrgyz government rapidly privatised the land used for agriculture, promoting the law on private property. Therefore the Soviet collective and state farms were abolished, the land divided among the farmers, and several farm enterprises, differing in size and crops, were established. One of the consequences of the reforms was that the number of land-share owners dramatically increased, leading to an urgent reforms package in the water sector.[225] In addition, due to the transition and to institutional lacks, disputes among farmers for water allocation and the deterioration of the canal network emerged. Hence, the water reforms developed in synchrony with the relatively early, rapid, and comprehensive land reforms. The Kirghiz government did not hesitate, contrary to other republics, to collaborate with international agencies; hence, under the influence of the WB and the ADB, the Kirghiz republic adopted a new strategy and issued legislation on water and irrigation management for introducing the WUAs. The first legal foundation of WUAs was adopted in the 1995 "Regulations of WUAs in rural areas" decree and in the 1997 "Statute of WUAs in rural areas"decree.[226] Since the mid-1990s, the control of most on-farm irrigation facilities shifted from governmental authorities to the newly established WUAs which started distributing water, maintaining the canals, and collecting the newly introduced fees. This practice provided the basis for charging the water users for all the WUAs' expenses without governmental subsides, with the aim to increase the farmers' responsibilities.[227] Despite of the donors' rationale supporting the WUAs based on hydrological boundaries, the

---

[224] WEGERICH, K., 2000. *Water Users Associations in Uzbekistan and Kirghizstan:Study on Conditions for Sustainable Development,* Occasional Paper n.32, Water Issue study group,SOAS, IWMI.

[225] WEGERICH, K., 2000. "cit.".

[226] GHAZOUANI, W. et Al, 2012. "cit."

[227] GUNCHINMAA, T., YAKUBOV, M., 2009."cit.".

Kirghiz associations were created based on the former collective farms' boundaries; most of the established WUAs (41 in 1999) lie in the Osh province, in the upstream Fergana Valley. Despite the introduction of the ISF and the restoration of part of the canals, it is questionable whether the total water amount for irrigation effectively decreased; the water quotas for the Kyrgyz republic in this transition phase were those determined by the 1992 Almaty agreement through the BVO Syr-Darja. Regarding the reform transition's progress, based on the decrees, in 2002 the government finally issued the Law on Water Users Associations, providing the basis for further development in irrigation management.[228] Although the Kyrgyz water sector showed significant changes at the on-farm level, due to the collaborative relationship between the government and the international donors, the Operation and Management (O&M) at the secondary/basin level was still under the supervision of governmental authorities. In contrast to Kyrgyzstan, in Tajikistan, the institutional reform path has been slow and problematic, affected by the civil war which lasted until 1994. Since the war, the land privatization process has gained pace and progressed quite quickly. The dismantling of the collective units led to a rise in private farms, though this phenomena did not significantly increase due to physical, political, and economic aspects; in Tajikistan, agricultural land is limited due to physical conditions, and in addition, after the USSR collapsed, an increase in pasture activities occurred. In the Tajik republic, the influence of international donors has been weaker, therefore, the water sector reforms proceeded slowly. The Water Code, adopted in 1993 and renewed in 2000, addressed some legal aspects related to the establishment of WUAs; the Code's rearrangement was supported by the WB, which started promoting the creation of Water Users Associations through international projectsin 1999. [229] Other WUAs were established through the support of non-government organizations (NGOs) without any official coordination with the Tajik government. Therefore, according to Sehring (2007), it was hard to determine the exact number of established WUAs in this country because only the associations supported by the WB and USAID were officially registered.[230] Regarding the WUAs economic aspects, since 1996 the ISF was introduced through a presidential decree; farmers have equally charged both those who rely on water lift irrigation and those who use gravity water. Hence, due to

---

[228] SEHRING, J, 2007. Irrigation Reforms in Kirghizstan and Tajikistan, *Irrigation Drainage Systems,* 21.

[229] SEHRING, J. 2007. "cit.".

[230] SEHRING, J. 2007. "cit.".

the physical features and the inevitably widespread use of pumping irrigation, water fees in Tajikistan are the most expensive in Central Asia.[231] Although water payment has been introduced, the unwillingness to pay and lack of money payments are still widespread among the farmers, primarily due to a cultural resistance, as previously analysed, and farmers' low income. Despite these issues, which were widespread in several water users associations, at the end of 2006, an official law regarding WUAs was finally approved by the government, in an effort to empower and formalize the farm-level water management actors. Nevertheless as Sehring claimed, a closer look at the established WUAs shows that they are performing differently than expected because they have not been established independently from local governance structures *(rayvodkhoz, former collective farms)* but instead are often dominated by them.[232] Whereas in Tajikistan the water reform process has been carried out — despite some problems, it was finally supported both by the government and the international donors — in Turkmenistan few changes have occurredsince the Soviet period. The irrigated area significantly increased to ensure food supply, mainly supporting wheat farming, but the reforms both in land and water management were somehow hampered by the government.[233] State and collective farms were dismantled, but despite some national reforms, the state control over land and water resources was maintained, in particular through state quotas for cotton farming and government subsidies to farmers, obstructing the farmers' efforts to carry out any activities related to the private market.[234] Although some informal WUAs were established, they are strictly controlled by the province and local water departments; furthermore, no official measures were issued by the government to formalize the Irrigation Management Transfer (IMT) to the local water users. Due to the Turkmen political and economic context, the international donors' actions have been weak and insignificant in the last decades, mostly hampered by the government's behaviour towards the international agencies. Focusing on Uzbekistan and Kazakhstan, in the last decades these two countries each carried out their own institutional water reform processes: whereas Uzbekistan has adopted a slowly paced water management reforms path, maintaining the state's prominent role in planned agriculture and water control, Kazakhstan's transition and

---

[231] GUNCHINMAA, T., YAKUBOV, M., 2009."cit.".

[232] SEHRING, J. 2007. "cit.".

[233] LERMAN, Z. & STANCHIN, I. 2004. Institutional Change in Turkmenistan' Agriculture: Impacts on Productivity and Rural Incomes, *Eurasian Geography and Economics,* 45, n.1.

[234] O'HARA, S, & HANNAN, T. 1999. "cit".

decentralization process hadalready been started at the time of the USSR collapse, and several international projects were created by the donors supporting
the IMT. These project were also established in Uzbekistan — such as the
IWRM in Fergana Valley which will be further discussed — even if the Uzbek
government showed an ambiguous attitude towards the international agencies.
These issues will be analysed in depth in the following chapters as they are the
core theme of this research. Since the 2000s, when the IWRM framework was
promoted worldwide by the water community, the international donors' actions
in the Central Asian republics have significantly increased. Their influence and
attention is mainly focused on the empowerment of the WUAs (institutional and
organizational tools) through a legal framework, and on the establishment of the
basin agencies, based on hydrological boundaries and related councils. To
achieve these objectives and complete the transition processes, the international
donors are supporting and financing the governments to design new national
legal frameworks formalizing the IWRM principles as the basis for the future
water management scenario. Although in some countries, such as Kirghizstan
and Kazakhstan, new water codes structured on the IWRM framework were issued, the evidence shows that the framework's implementation has not been
completed, in some cases influenced and hampered by local and provincial political powers. Therefore the IWRM implementation still represents a great challenge for water resources management in Central Asia. Focusing on Uzbekistan
and Kazakhstan, both at the national and local levels, these issues will be analysed in depth and compared in the following chapters.

# 4. THE WATER REFORMS AT THE BASIN LEVEL: COMPARING THE MIDDLE ZERAVSHAN VALLEY (UZB) AND THE ARYS VALLEY (KAZ)

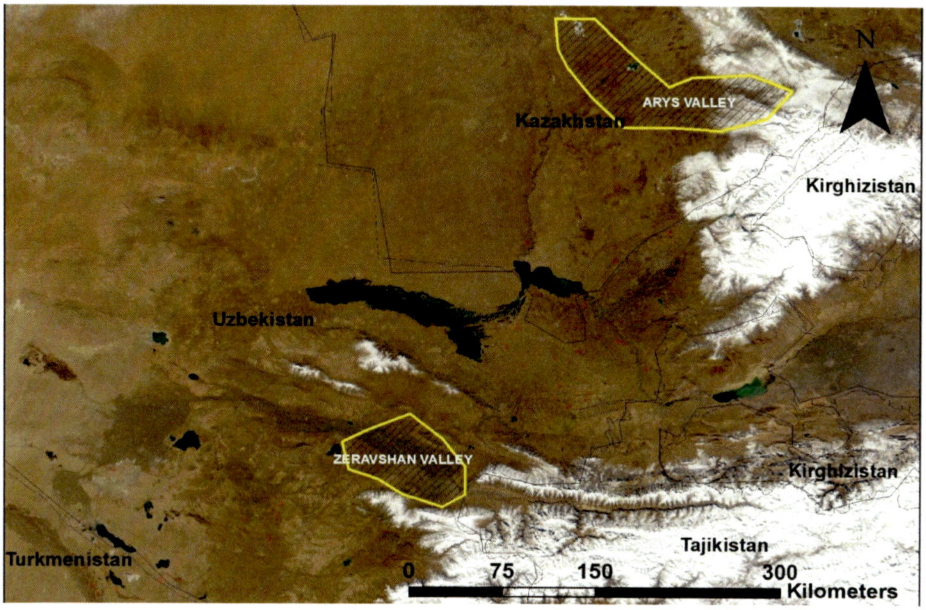

*FIG. 9: GIS elaboration of a satellite image (NASA-Modis, 2003) representing the two regions (in yellow shapes) in Uzbekistan and Kazakhstan chosen as field-research areas.*

## 4.1 TOWARDS THE IWRM IN UZBEKISTAN AND KAZAKHSTAN: EVIDENCE FROM THE NATIONAL LEVEL

As analysed in the other Central Asian republics in Chapter 3, this section focuses on the national water reforms path carried out in the independent nations of Uzbekistan and Kazakhstan since the collapse of the Soviet Union. As it has been described in part already, in the post-Soviet Aral Sea basin territory, although relevant new water institutions were created (ICWC and IFAS), most of the inter-republican water agreements did not change after 1991, but instead were maintained by the newly independent countries. The downstream countries, such as Uzbekistan and Kazakhstan, had no interest in renegotiating the water quotas while the upstream ones, Kirghizstan and Tajikistan, were too weak to push through a change to the regional agreements. At the national level, the institutional changes that affected all the governments' apparatuses inevitably led to consequences for the water sector and its institutional framework. Though Uzbekistan has kept a political state-centralized approach in natural resources management, Kazakhstan has been more oriented towards a decentralization process and has adopted formal changes towards the market economy. Nevertheless, focusing on the politics of water reforms, several factors were identified which are influential and shared by both Uzbekistan and Kazakhstan: the economic development and structure, water resources and water use, historical water management institutions, national policy priorities, and the state of financial and technical capacities. These aspects are essential for the understanding of the water reform context in the two countries.[235] The economic development of the two republics is strictly connected with the maintenance of water facilities, funding for research and new infrastructures, and payment for experts. Furthermore, the importance of irrigated agriculture must be considered, which is the most water-consuming activity worldwide, in developing countries' economies; focusing on GNP, irrigated agriculture represents a relevant sector both in Uzbekistan and Kazakhstan, although with differences. Even though in the last decade Kazakhstan oriented its economy towards the exploitation of oil and gas resources and its international exports, the percentage of agricultural production in the national GNP is still 8–10%; the percentage is more in South-Kazakhstan where 60–70% of the irrigated areas are located, while in Uzbekistan it rises to 28%–30% of the national GNP. [236] This high rate of agricultural production in the Uzbek economy is mostly due to cotton farming, which is especially water-intense, and to wheat and rice farming. The total irrigated area in Uzbekistan reaches 4.2 mil. hectares, while in Kazakhstan it reaches 1.3 mil. hectares, mostly located in southern provinces, such as South-Kazakhstan, Djambul, Almaty, and Kizylorda.[237] In addition, subsistence agriculture has become increasingly important, in particular for populations who live in rural areas, with 65% in Uzbekistan and 46% in Kazakhstan.[238] Nevertheless, the water sector represents a key role in both countries' economies, although with differences and contrasts, being strictly connected with political, social, and environmental issues. Although Uzbekistan and Kazakhstan are crossed by the most important Aral

[235] SEHRING, J., 2006. "cit.".
[236] WORLD BANK, 2012. *World Bank Development Indicators*".
[237] DUKHOVNY V., DE SCHUTTER J., 2011. "cit.".
[238] WORLD BANK, 2012. "cit.".

Sea basin rivers — that is, Amu-Darja, Syr-Darja and Zeravshan — most of their territory lies in arid and semi-arid environments featured by annual precipitation ranging from 100 mm/y in the west downstream areas to 400–500 mm/y in the foothill areas, where most of the cities are located.[239] Irrigated agriculture is widespread along the rivers and in the irrigation schemes from the eastern foot-hill areas to the western downstream areas in both countries. The similarities in geographical features and agricultural parameters resulted in similar water use patterns both in Uzbekistan and Kazakhstan. As previously analysed, the two countries use more than the 60% of the total Syr-Darja flow, according to the ICWC–BVO Syr-Darja quotas, and the general conditions of the water facilities and infrastructures in both the countries are similar, both having been built by the Soviet Union during the 1960s to the end of the 1980s; the economic decline following the USSR's collapse led to the dysfunction of part of the large irriga-tion systems.[240] Though the irrigated area in Uzbekistan has not decreased since independence, in Kazakhstan it decreased throughout 2 mil. ha to the present 1.3 mil. ha; this decrease was due to the dismantling of the pumping stations, de-cided and carried out by the Kazakh government since 1992.[241] It must be em-phasized that in both countries the institutional water reforms have been strictly related to those of the agricultural sector, meaning the dismantling process of the state and collective farms, *sovkhoz* and *kolkhoz*.

### *4.1.1 The water management structure in Uzbekistan*

During the 1990s Uzbekistan carried out a complicated and challenging transi-tional path oriented towards decentralization of powers within state authorities rather than effective openness towards market principles, both in the water and agricultural sectors.[242] Focusing on the water sector, in the first years of the 1990s, the Uzbek institutional water framework and its management structure and actors had not significantly changed from the Soviet era. On May 6th, 1993, the first legislation on water use since independence, the "Law on Water and Water Use", was enacted by the Cabinet of Ministers of Uzbekistan which is responsible for its implementation.

---

[239] SUSLOV, S.P., 1961. "cit.".

[240] DUKHOVNY, V., DE SCHUTTER, J., 2001. "cit.".

[241] POMFRET, R., 2007. Rebuilding Kazakhstan's Agriculture, *Central Asia and Caucasus Analyst.*

[242] KANDIYOTI, D., 2002. *Agrarian Reforms, Gender and Land Rights,* Social Policy and Development Programme Paper n.11, United Nations Research Institute for Social Development.

According to the 1ˢᵗ Article, water legislation tasks are the following: [243]

*"Water relations regulation, rational water use for needs of population and economy, water protection from pollution and exhaustion, water harmful impact prevention and mitigation, water structures improvement, enterprises and organizations' water rights protection".*

Nevertheless, this measure did not lead to significant changes and did not introduce any new patterns for water management and use; water was managed, as during the Soviet Union, according to administrative boundaries from the national to the farm level. The Ministry of Melioration and Water Resources *(Minvodkhoz)* had its own department at the province *(Oblastvodkhoz)* and district levels *(Rayvodkhoz)*; the district water department was responsible at the local level for the water supply to the farms. [244] This structure was altered just at the national level in 1997 when the Ministry was merged with the Ministry of Agriculture and a new Ministry of Agriculture and Water Resources (MAWR) created. The legal water framework adopted in 1993 did not lead to any significant changes because in the first years of independence, the national agricultural system also had not been affected by significant changes. The state-oriented governmental approach mentioned above was partly due to the state quotas for cotton and wheat, which were maintained, in contrast to the other Central Asian countries, except Turkmenistan and Tajikistan (only cotton), which had been part of the Soviet system. Therefore, a liberalisation according to the market principles have not been supported and an open market for these crops among the farmers have not been institutionalised; in addition the agricultural lands were kept under the state property. [245]

### 4.1.2 The de-collectivization of the Uzbek land and agriculture

In Uzbekistan in 1991 the agricultural land was under the shared control of 971 *kolkhoz* and 1137 *sovkhoz*; starting from 1993 the state and collective farms were dismantled and transformed into producers' cooperatives named *shirkat*.

---

[243] DJALALOV, A.A., 2001. *National Water Law of Uzbekistan: Its Coordination with International Water Law. Priorities and Problems. Line of Activities for Improvement,* ICWC Training Centre Seminar "International and National Water Law and Policy, September 24-29, 2001.

[244] WEGERICH, K., 2006. "cit.".

[245] BOCH, P., 2002. *Agrarian Reforms in Uzbekistan and in other Central Asian Countries,* working paper n.49,Land Tenure Centre, University of Wisconsin, Madison.

Those new typologies of farms did not present several differences in terms of land rights, water supply, and organizational and territorial structure compared to the Soviet collective farms; most of them were still managed by the apparatus which worked during the Soviet Union and most of the employees — called *bri-gadi* in the *kolkhoz* and renamed *pudrats* — went on working on the farms. The *shirkats'* total land unit was reduced from an average of 5000 to 6000 ha for a single collective farm to 2000 ha, but the farms continued to grow mostly cotton, wheat, and rice.[246] It is important to emphasize that wheat farming in Uzbekistan had been introduced since independence in order to achieve national food self-sufficiency because of the end of import trade flows from the other Soviet republics. Therefore, thousands of hectares of irrigated land all over Uzbekistan, in particular in the upstream foothill areas, were converted from cotton to wheat farming; the total area for cotton farming decreased of 30–35%.[247] Wheat national self-sufficiency had already been achieved by 1994, and in addition since 2000 an export flow to neighbouring countries, such as Afghanistan and Tajikistan, had been initiated. Although wheat farming is also included in the state-quota system, it is somewhat more flexible than cotton farming: the farmers are obliged to sell 50% of their wheat output to the state.[248] At the end of the 1990s the Uzbek government, while trying to increase agricultural productivity and following the ongoing policies of the neighbouring republics, issued a significant measure with the aim of changing the agricultural structure at the farm level; in April 1998 a new land reform named "Law on the Peasant Farmers" issued by the Cabinet of Ministries of Uzbekistan introduced the progressive dismantling of the cooperatives, *shirkats,* and opened the possibility for farmers to rent agricultural units from districts governments.[249] Farmers could sign a land-leasing contract for an agricultural plot, often part of the *shirkat* land where they previously had worked, ranging on average from 5 to 15 ha, depending on the crops, for 49-year periods.[250] Although this measure promoted private farming and land tenure liberalisation, it was a promotion *de facto* because it did not lead to any changes to the Uzbek state agricultural system. The cropping

---

[246] ILKHAMOV, A., 1998. Shirkat, Dekhqon farmers and others: Farm restructuring in Uzbekistan, *Central Asian Survey,* 17:4.

[247] SPOOR, M., 2006. Uzbekistan' s Agrarian Transition, in: Chandra Suresh Babu and Djalalov S. (eds.) *Policy Reform and Agricultural Development in Central Asia,* Boston, Springler.

[248] AMINOVA, M., ABDULLAEV, I., 2009. "cit.".

[249] KANDIYOTI, D., 2002. "cit.".

[250] DUKHOVNY, V., DE SCHUTTER, J., 2011. "cit.".

plans were still organized and issued by the state through the state quotas for cotton and wheat farming. The regional and district departments provided to the farmers all the agricultural inputs for these crops, such as fertilizers and machinery, therefore obliging the farmers to fulfil the plan signed with the province government (*hokimyat*). In addition, according to Trevisani (2007), the farmers' association (Fermer and Dekqon Association–FDA) was established at the republican level with provincial and district branches, with the declared aim of supporting the development and the strengthening of the private farms, *de facto* sidelines the district department of agriculture in the coordination of the state crops planning.[251] The effective and widespread implementation of the 1998 "Law on Peasant Farmers", and the subsequent farm-restructuring process, required several years due to the significant sociopolitical changes involved in transitioning from the collective unit to the private one; although the creation of private farms speeded up starting from 2002, in that year at least the 65% of Uzbek agricultural land was still controlled and managed by the *shirkats*.[252] The following year, in 2003, the government enacted a measure regarding the dismantling of all the cooperatives which were still working in the country. Despite this measure in the Middle Zeravshan valley, one of the most important irrigated areas in Uzbekistan, in 2005 private farms represented on average only 60% of the agricultural farms.[253] At the national level over 600 *shirkats* were dismantled during 1999–2004 and by early 2006 the remaining ones were also abolished. The elimination of all the remaining cooperatives in 2007 marked the end of collective farming in Uzbekistan and paved the way for the full introduction of individual private farming systems throughout the country. Private farms increased from about 23.000 in 2003 to 141.000 in 2007, cultivating about 75% of agricultural land. According to Yalcin and Mollinga (2007), one of the main reasons that led the government to encourage this major change to private farming was the increasing difficulty of policing cotton and wheat production; furthermore, government officials estimated that production costs on private farms was 20% or even 30% less than they had been on *shirkats*. In addition, it seemed that the government had also overcome its earlier reluctance and opposition to individual farms.[254] This structural change in the agricultural system led to great

[251] TREVISANI, T., 2007. After the Kolchoz: rural elites in competition, *Central Asian Survey,* 26:1.

[252] SPOOR, M., 2003. "cit.".

[253] ZINZANI, A., 2011. Tra Irriguo e Seccagno: le Trasformazioni  Recenti nella Media Valle dello Zeravshan, Uzbekistan, *Rivista Geografica Italiana,* 118.

[254] YALCIN R., MOLLINGA, P., 2007. *Water Users Associations in Uzbekistan: the Introduction of a*

challenges and a need for change in water resources management, that remained the same, except for the Law on Water (1993), since the USSR's collapse. The breakdown of the cooperatives led to the emergence of thousands of independent farmers; therefore, this phenomenon constituted the need to devise some mechanisms for water distribution over smaller units than before, when the *shirkat* was the unit to which government managers supplied water. District water departments, were no longer able to provide water to thousands of new private farmers, as they were formed to provide planned water supply to the collective units. In addition, the irrigation schemes were designed during the Soviet Union for this typology of units and not for private farms. In addition to these structural agricultural changes oriented to the privatisation of land management, since the end of the 1990s governmental ideas and policies have been partly influenced by the action of the international donors and funding agencies.

### 4.1.3 The controversial reform process of the national water sector

Yalcin and Mollinga (2007) stress that successful structural changes were achieved through the interaction of social and political networks within the government bureaucracy supported by international agencies.[255] In fact, as it was mentioned in the previous chapter, for the whole Central Asian region, including Uzbekistan, since the end of the 1990s the international donors, specifically the SDC, USAID, and the WB, have started promoting a new water management rationale based on the backbone of the IWRM framework.[256] In regard to the changes in the agricultural context, the donors and funding agencies initiated promoting the Irrigation Management Transfer (IMT) from the governmental departments to the water users through the establishment of the WUAs. The institutional and organizational backbone of these associations included some of the main pillars of the IWRM framework — such as water organizations based on hydrographic boundaries, the introduction of irrigation service fees (ISF), the integration and participation of the water users through the support of efficient governance in water decision-making processes.[257] Moreover, at the basin level, the establishment of basin agencies for water control and supply — replacing the

---

*New Institutional Arrangement for Local Water Management,* Amu Darja Case Study- Uzbekistan, Deliverable WP 1.2.3 of the NeWater project, Centre for Development Research, University of Bonn.

[255] YALCIN R, MOLLINGA, P., 2007. "cit.".

[256] DUKHOVNY, V.V., SOKOLOV, V.V., 2006. "cit.".

[257] SEHRING, J., 2007. "cit.".

province water departments (*oblastvodkhoz*) — was supported, according to the same principles. The focus on the water sector reforms at the basin/meso level in Uzbekistan will be analysed in depth in the next section. Although, on the one hand, relevant water reforms at the basin/local level have been carried out since the 2000s, on the other hand, at the national level no significant measures were enacted. In 2009 some amendments were added to the "Law on Water Use" of 1993; their implementation at the basin/local level, will be further pointed out. In contrast to some neighbouring countries, such as Kirghizstan and Kazakhstan, in Uzbekistan new national water codes have not been enacted since the USSR's collapse, limiting significant changes in water resources management at the national level. It is important to state that whereas the international donors were able to introduce some relevant reforms concerning the establishment of the basin agencies and the WUAs, at the national level, the support of the government, regarding the new code, have not led to any concrete results yet. Therefore, although the IWRM's pillars have been supported and partly implemented through the issued measures at the basin level and the donors-based projects, the framework has not been institutionalised yet. This institutional lack keeps the IWRM's implementation weaker and more difficult in comparison with some of the neighbouring countries. According to representatives of Central Asia Scientific Research Institute for Irrigation (SANIIRI) and the German Society for International Cooperation (GIZ), interviewed as national-level experts during the first round of fieldwork in 2011, a governmental debate on the issue of a new water code, supported by the international donors, is ongoing; both institutions claimed that possibly a new legal water framework, officially supporting the IWRM, could be potentially enacted in 2015/2016, though they did not affirm any certainty about this process.[258] The reasons behind this resistance can be found by analysing the swinging relationships among the government and some international donors; in the last years some of the funding agencies operating in the Central Asian region and in Uzbekistan have criticized the government because of its policies on water resources — that is, for being still state-oriented — and in particular on agricultural processes (this issue will be further discussed later). State quotas for cotton and wheat (50% of the yield) farming are still operative, limiting the crop choices and the free market among farmers. Furthermore, according to Abdullaev (2009), the cotton and wheat quota system has a negative impact on water management and use; water management organiza-

---

[258] Personal communication with SANIIRI and GIZ, Tashkent (Uzbekistan), April 2011.

tions are forced to deliver water to thefarms that grow state-quota crops first, and withhold supplies from potentially higher-value agricultural users.[259]

In addition, state quotas shape water governance processes, because planning, distribution, and control are done through state water management organizations, and since state quotas are a part of the overall state policy, they predetermine a participated water governance.[260] The Centre for Economic Research (CER) experts added that, despite the recent establishment of donors' projects allowed by the national governmental institutions, it will be hard nowadays to have a stable context in which to design a new water legal framework collaborating with the international agencies.[261] Anyhow, if the national reform process in water management will go on according to the IWRM framework, it will be necessary to strengthen the newly established agencies.

### 4.1.4 The water management structure in Kazakhstan

In Kazakhstan evidence has shown a different transition path both in the water and agricultural sectorsin the last decades — that is,a path oriented more towards a liberalisation and decentralisation process as well as towards the introduction of market principles in the national economy. Although, as mentioned above, the geographical and water availability features are quite similar compared to Uzbekistan, the political and, in particular, the economic, context have presented significant differences.[262] As briefly described above, after independence and in particular during the 1990s, at the national level, governmental action has focused more on natural resources management and exploitation of oil and gas, which lie mostly in the Caspian Sea, rather than on agricultural and water resources. According to Pomfret (1997), the general policy stance towards agriculture was one of neglect as the ministers focused on macroeconomic stabilization, privatisation, and the development of the petroleum sector.[263] Therefore, as a consequence of this national economic trend, the agricultural percentage in the GNP decreased from 15% in 1991 to the 8–10% of 2005. As it was mentioned above, in the South-Kazakhstan province this decrease has been less relevant, and nowadays the agricultural percentage reaches more than 15% of

---

[259] ABDULLAEV, I. et Al., 2009. Agricultural Water Use and Trade in Uzbekistan: Situation and Potential Impacts of Market Liberalisation, *Water Resources Development,* vol. 25, n. 1.

[260] AMINOVA, M., ABDULLAEV, I., 2009. "cit.".

[261] Personal communication with CER, Tashkent (Uzbekistan), April 2011.

[262] DUKHOVNY V., DE SCHUTTER, J., 2011. "cit.".

[263] POMFRET, R., 2007. "cit.".

GNP; these different economical conditions are due to the lack of oil and gas resources in this territory and to different climate conditions and water resources availability, compared to central-northern Kazakhstan, which better allows agricultural practices. However, although the national irrigated area decreased from 2 mil. ha to 1.3 mil. ha after the USSR's collapse due to the end of state subsidies for pumping stations' maintenance, in South-Kazakhstan the decrease of cropping land was not significant. Concerning the national water institutional structure, contrary to Uzbekistan that kept the Ministry of Water Resources and Melioration (*Minvodkhoz*) within the new legislation after independence, in Kazakhstan the Soviet *Minvodkhoz* was not replaced by a new ministry but was transformed into the State Committee of Water Resources (CWR) under the Cabinet of the Republic of Kazakhstan.[264] After independence Kazakhstan was the only country without a Ministry of Water Resources in the Central Asian region. Therefore, according to Burger (1998), the role of Kazakhstan in international relations based on water resources has diminished with the establishment of the state committee compared to the other Aral Sea basin republics.[265] In March 1993 the water sector in Kazakhstan was regulated by the "Water Code", issued by the Cabinet of Ministers, which describes management principles and responsibilities with regard to water management. The following lines summarize the Code's aims [266]:

---

[264] ZIMINA, L., 2003. Developing Water Management in SouthKazakhstan, in *Drop by Drob: Water Managment in the Southern Caucasus and Central Asia,* edited by S. O'hara, Local Government and Public Service Reform Initiative, LGI Fellowship Series, Open Society Institute.

[265] BURGER, R., 1998. *Water Users Associations in Kazakhstan: an Institutional Analysis*, NIS PROJECT, Harvard Institute for International Development, Environment Discussion Paper n.45.

[266] RAMAZANOV, A.M., 2001. *National Water Law of Kazakhstan:Its Coordination with International Water Law. Priorities and Problems. Line of Activities for Improvement,* ICWC Training Centre Seminar "International and National Water Law and Policy, September 24-29, 2001

[266] ZIMINA, L., 2003. Developing Water Management in SouthKazakhstan, in *Drop by Drob: Water Managment in the Southern Caucasus and Central Asia,* edited by S. O'hara, Local Government and Public Service Reform Initiative, LGI Fellowship Series, Open Society Institute.

[266] BURGER, R., 1998. *Water Users Associations in Kazakhstan: an Institutional Analysis*, NIS PROJECT, Harvard Institute for International Development, Environment Discussion Paper n.45.

[266] RAMAZANOV, A.M., 2001. *National Water Law of Kazakhstan:Its Coordination with International Water Law. Priorities and Problems. Line of Activities for Improvement,* ICWC Training Centre Seminar "International and National Water Law and Policy, September 24-29, 2001.

*"The Water Code establishes a legal base of rational water use for population needs, economic branches and environment, water resource protection from pollution and exhaustion and harmful water impact elimination"*

According to Item 1 Article 11 of the Code, water resources management is executed by the government, local authorities, and state water agencies within their competence; in addition, Article 47 affirms that water use is executed free of charge. Debating the Water Code, Wegerich (2008) supposed that the new legal framework would open the way for the introduction of a market economy in irrigated agriculture and allow for the establishment of WUAs, despite the fact that the law affirms that water management is up to the government.[267] According to Zimina (2003), as early as 1992, the government introduced a system of water pricing, but only in 1997 were tariffs introduced to specify charges for different sub-sectors in each river basin.[268]

### 4.1.5 The land tenure national reforms

Focusing on the agricultural land sector, the reform processes were carried out from 1993 to 1998 and can be divided into two phases; emphasis was placed on a restructuring of the former state and collective farms into cooperatives and large collective enterprises, not so different from the first step of reforms that occurred in Uzbekistan.[269] In early 1996 the privatization process was already almost completed; 93% of the country's state farms (*sovkhoz)* had been privatised and all the former collective farms (*kolkhoz)* had been reregistered as private entities.[270] The reform process was completed through the issue of the Land Code by the Cabinet of Ministers in late 1995; although the Kazakh agricultural land still remained state property, the private farmers — most of them employees of the former collective farms — were allowed to lease land from the state on a long-term basis (99 years).[271] The average plot ranged from 5 to 15 ha, depending on the quality of the land and other factors, and it could be rented to other farmers, but the sale of agricultural units was forbidden. Strengthening the reform path toward market economy principles, the state quota for crops, cotton,

---

[267] WEGERICH, K., 2008. Blueprints for water users associations' accountability versus local reality: evidence from South Kazakhstan, *Water International,* vol.33, n.1.

[268] ZIMINA, L., 2003. "cit.".

[269] WEGERICH, K., 2008. "cit.".

[270] BURGER, R., 1998. "cit.".

[271] POMFRET, R., 2007. "cit.".

wheat, and rice was abolished in 1995–1996; only a small percentage of wheat (on average 20%) was still purchased by the government. Therefore, in the middle of the 1990s the Kazakh farmers were free to cultivate and sell their own output at market prices; regarding these aspects, it is evident how this early reform step differs from the one conducted in Uzbekistan. The second step of the agricultural reform occurred in 1998 through the issue of the bankruptcy law which defined the practical application of bankruptcy to the farm sector; the large cooperative enterprises were the farms which were dismantled enhancing the rise of private farmers. According to Wegerich (2008), by April 1999 already 85.000 peasant farm entities werelegally recognized, but due to the difficult registration process the number of formal and informal farms was estimated to be approximately 200.000–250.000.[272]

### 4.1.6 A water reforms process oriented to the IWRM

As occurred in Uzbekistan, land and agricultural reforms required a similar transition path to the water sector's structure and legislation. At the end of the 1990s, significant changes affected the national water management's structure: the State Committee for Water Resources, established after independence, was transformed into the Committee for Water Resources within the Ministry for Natural Resources and Environmental Protection and in 1997 the Committee was transferred to the Ministry of Agriculture.[273] These changes aimed to raise the profile of irrigation water consumption and consequently of agriculture; the national water structure was finally stabilized in 2002. Despite these institutional changes, since 1993, according to the Water Code, the Committee for Water Resources has basin branches offices all over Kazakhstan: specifically, eight branches based in the basin unit. These Committee branches are working in collaboration with the republic, province and district water departments; the water institutional structure at the basin/local level will be deeply analysed in following sections. As already occurred in Uzbekistan, the emergence of thousands of new private farmers at the end of the 1990s led to a debate at the governmental level concerning the Irrigation Management Transfer (IMT) of water facilities and services to the water users; these processes have been supported by the international donors — in particular by the WB, ADB, USAID and UNDP — operating in different ways in Kazakhstan since the second half of the 1990s, due

---

[272] WEGERICH, K., 2008. "cit.".
[273] ZIMINA, L.,2003. "cit.".

to the more open and accessible political environment in comparison with Uzbekistan. Besides the support of the WUAs' establishment, the international agencies, through their collaboration with the government and the creation of development projects, induced the introduction of the IWRM's pillars in the water institutions. As a first step, despite a system of water pricing that had been already created in 1992, in 1997 water tariffs were introduced for the water users; the water charge was supposed to be calculated in cubic metres, but due to the obsolete conditions of the water facilities built for large collective farms, water distribution was estimated rather than correctly calculated.[274] Furthermore, the farmers had difficulty managing the Irrigation Service Fee (ISF) due to their lack of experience and coordination as well as lack of participation in the water delivery procedures . Despite some problems due to the inexperience of the actors involved in these institutional changes, the WUAs were established (and will be further analysed) and the implementation of the IWRM's pillars, supported by the donors, gained significance in the national governmental institutions. In January 2002 the government of Kazakhstan passed a resolution to approve  the water sector development and water policy in the republic until the year 2010; the main goal of this initiative was to define the main actions for conservation and efficient use of water resources. The initiative was used as a basis to improve the legal framework in Kazakhstan for developing water programs and actions.[275] Recognizing the inadequacy of the 1993 Water Code to strengthen the reforms in the water and agricultural sector conducted since 1996, as also claimed by Zimina in 2003, the government of Kazakhstan, with the support of the UNDP, enacted a new Water Code in July 2003; through this measure the IWRM's framework was institutionalised as the strategic and challenging pattern for future development in the water and environmental sector. The new Water Code's main goals for the IWRM implementation in Kazakhstan were the strengthening and the empowerment of the  eight  River Basin Agencies *(BWO)* already operating within the Committee of Water Resources (CWR), and in particular, the establishment of the River Basin Councils (RBC).[276] The Basin Councils' creation would be the basis to shift from pure basin water management to basin water management and governance, supporting the integration

---

[274] WEGERICH, K., 2008. "cit.".

[275] GOVERNMENT OF R.K. /UNDP, 2004.*National Integrated Water Resources Management and Efficiency Plan in Kazakhstan,* Project document.

[276] UNDP-GOVERNMENT OF NORWAY, 2005.*Kazakhstan National Integrated Water Resource Management and Efficency Plan,* draft of a project document.

and the participation of the water users in the decision-making processes. Furthermore, with the main aim of institutionalising the reforms at the basin/local level, in 2003 the Law on Water Users Associations (Law n.404-II) was enacted by the Cabinet of Ministers; this measure, which will be specifically highlighted later, has formalized the WUAs' institutional and organizational role as a rural cooperative of water users (*SPKV*).[277] The implementation of the IWRM's pillars through the enacted Water Code represented a great challenge for the Kazakh government and for the Committee of Water Resources. Therefore, in order to support and facilitate the ongoing process, in 2004 the CWR was assisted in the preparation of the "Kazakhstan National Integrated Water Resources Management and Efficiency Plan" planned by the UNDP through the project for a "National IWRM and Efficiency Plan for Kazakhstan" for the period 2004–2008. The project was funded by the government of Norway and partly by the UK Department for International Development; from the regional perspective, Kazakhstan has been the first national IWRM and Water Efficiency Plan project in Central Asia and, indeed, in the CIS region as a whole.[278] According to the international donors and the Kazakh government, the Committee for Water Resources was appointed as an implementing agency. The main goals of the project were the empowerment of the River Basin Agencies, which had been already established but their role needed to be strengthened, and, in particular, clarifying their aims towards their partner institution, the Republican State Enterprises (*RGP*). In addition, working in collaboration with the BWO, the creation of the River Basin Councils was supported to establish a governance structure. This process represented a great challenge for Kazakh basin units, as it required the involvement of all the water users — from the farmers, to the WUAs and District Water Departments' members. The following is a scheme of the current water management structure:

---

[277] Personal Communication with GIZ, Tashkent, April 2012, with Research Institute on Water Economy, Taraz, April 2012, with South-Kazakhstan Hydrogeological State Enterprise, Shymkent, November 2012.

[278] GOVERNMENT OF R.K. / UNDP, 2004."cit.".

| National Committee of water resources (Ministry of Agriculture) |
| :---: |
| River basin agencies (BWO) – branches of Committee $\rightarrow$ River Basin Councils |
| Republic State Enterprises (RGP) – Province level |
| WUAs (SPKV) - District Water Departments (*Kommunalnivodkhoz*) |
| Farmers |

As it will be discussed in the next sections, the establishment process of the River Basin Councils that began in 2005, in recent years has not been easy and immediate, being partly hampered and not supported by the local actors, mostly due to a lack of competencies and knowledge regarding the promoted IWRM principles.[279] However, according to the project partners, the capacity building that was achieved allowed for an increase of the staff, of the organizational and technical support, and the training of the WUAs' members. In 2008 the "National IWRM and Efficiency Plan for Kazakhstan" project was extended until 2025 and divided into two phases, 2008–2010 and 2010–2025; the objectives of the project's extension have not changed since the initial step in 2004. Although, according to the project partners, the river basin councils were established at the end of 2008, the project went on to reinforce the new institutions, which were still unstable and weak concerning several aspects.[280] Several issues in the last years have kept the IWRM's implementation process slower, more problematic, and partially blocked, hindering it more than was expected when the project was established. The project's partners claimed that, although a fair and correct Water Code was enacted, the Kazakh political background, and the subsequent organizational immaturity and sectoral fragmentation, partly hampered the process and made the course difficult. Furthermore, they mentioned that although the organizational water reforms were recognized at a high national level, no actual reforms were effectively put in place, in particular at the local level. Also, the low status of the River Basin Agencies, the bureaucracy, and lack of organization and efficiency could be a threat to the project's development.[281] Evidence from the basin/local level and interviews with the related

---

[279] Personal communication with GIZ and IWMI, Tashkent, April and October 2012.

[280] UNDP, 2007. Programme "Integrated Water Resource Management and Water Efficiency in the Republic of Kazakhstan for 2008-2025", Progress Report, Undp project #39257.

[281] UNDP, 2007."cit.".

experts, which will be further highlighted, showed an unstable context in the Kazakh national water sector and a possible future turnaround in the reforms' implementation process. The UNDP members claimed that the low status and the subordinated nature of the Committee of Water Resources, which is in the organizational structure of the Ministry of Agriculture, does not promote efficient implementation of state policies about use and protection of water resources, inter-sectoral coordination, and the sustainable development of the water sector. In addition, the Ministry position in international relations and agreements results weaker in comparison with the neighbouring countries, in particular with Uzbekistan. As emerged from the interviews with representatives from both national institutions and international agencies, for some years a debate has been going on at the governmental level about a change in the institutional framework of the Committee of Water Resources.[282] The change would lead to a stronger position of the national water authority both in the national and international sector. According to the interviewed experts, three main options are under-evaluation: one option could be to modify the structure and the status of the actual Ministry of Agriculture, renaming it "Ministry of Agriculture and Water Resources", while another option is to focus on the creation of a Ministry *ex-novo* of Water Resources; a third option is to establish a state agency of water resources directly under the control of the prime minister.[283] Although these options have been under discussion for three to four years, at present there is not any certainty about the future institutional and operational assets of the national water authority.

---

[282] Personal communication with Research Institute on Water Economy, Taraz, April 2012.

[283] Personal communication with Research Institute on Water Economy, Taraz, April 2012, GIZ and IWMI, Tashkent, April and November 2012, and with South-Kazakhstan RGP, April 2012.

## 4.2 THE WATER REFORMS PROCESS AT THE BASIN / LOCAL LEVEL IN UZBEKISTAN

In the previous section the focus has been on the water sector reforms at the national level in Uzbekistan; some elements of these themes at the basin/local level have already been anticipated, mainly focusing on the establishment of new water institutions according to the IWRM and IMT rationale, basin agencies, and WUAs, replacing the province and district water departments (*oblastvodkhoz* and *rayvodkhoz)* inherited from the Soviet Union. This section focuses on the institutional path which led to the establishment of these new entities of basin level water management. As it was briefly mentioned above, since the end of the 1990s, the international donors working in the Central Asian region started promoting the development of the water sector based on the IWRM's principles; specifically they supported the Irrigation Management Transfer (IMT), the entities territorially based on hydrographic principles, the introduction of the Irrigation Service Fee (ISF), and the creation of a governance structure based on the participation of the water users in the decision-making processes.[284] In Uzbekistan the international donors and funding agencies which started collaborating with the Uzbek government with the aim of introducing new ideas and strengthening the reforms according to the IWRM principles have mostly been the WB, the ADB, the UNDP, and the USAID. According to Yalcin and Mollinga (2007), the ideas for the water sector changes have entered into the Uzbek governmental system from outside via the international organizations; playing a crucial role in the reforms' initiatives, the ideas have taken time to materialize,be absorbed into the bureaucratic system, and then be explained in a language which fits the local political culture.[285]

### 4.2.1 The conflicting move towards hydrographic water management at the basin level

Since the 2001 initiation of the agricultural transition process, the establishment of the basin agencies and the WUAs aimed at filling the institutional gap that had emerged in the on-farm irrigation systems after the dismantling of the collective farms which used to be responsible for these services and improving the efficiency of water resources usage at the basin / local level.[286] Therefore, in 2001 a special commission formed by the Cabinet of Ministers of the Republic

---

[284] DUKHOVNY V.V., DE SCHUTTER, R., 2011. "cit.".
[285] YALCIN, R., MOLLINGA, P., 2007. "cit.".
[286] GUNCHINMAA, T., YAKUBOV, M., 2009."cit.".

of Uzbekistan prepared a "Program of measures on the improvement of irrigated lands for 2001–2010" based on a two-level scheme for basin agencies and WUAs. The first step, focusing on the basin units, proposed the transition of water management and control from the province water departments *(Oblast-vodkhoz)*, based on administrative principles according to the USSR's rationale, to seven basin water administrations according to hydrological principles; this measure would be the first step, at the basin level, oriented to the IWRM principles' implementation.[287] This important institutional change was driven and promoted by Abdurakhim Djalalov, the Minister of Agriculture and Water Resources from 1999 to 2004; as stated by some of the interviewed Uzbek experts, the former Minister worked in very close collaboration with the international donors and funding agencies operating in Uzbekistan and participated in several conferences organized by the international water community.[288] As also mentioned by Yalcin and Mollinga (2007), Djalalov argued that Uzbekistan and some other Central Asian countries were, at the end of the 1990s, part of the few countries in the world which were still managing water according to administrative principles, in spite of the ideas of the international water community; therefore, he argued that, because of the intensification of structural reforms both in the water and agricultural sectors, it was necessary to review the existing national water management structure.[289] The reform was oriented to the shift from the fourteen Uzbek Province Water Departments — one for each province and one for Tashkent city, to seven water basin agencies based on *hydrographic* principles, putting together different provinces under the control of one agency; here follows the list of the promoted new authorities (FIG.10) [290]:

---

[287] IWMI, 2012.*Review on WUAs' development in Uzbekistan,* unpublished report.

[288] Personal communication with IWMI and GIZ, Tashkent, October 2012.

[289] Personal communication with IWMI and GIZ, Tashkent, October 2012 / YALCIN, R., MOLLINGA, P., 2007."cit.".

[290] IWMI, 2012."cit.".

| | |
|---|---|
| Fergana Valley Basin Agency 906.000 ha | Andijan, Fergana, Namangan Provinces |
| Chirchik Basin Agency 396.000 ha | Tashkent Province |
| Syr-Darja Basin Agency 594.000 ha | Syr-Darja and Jizzakh Provinces |
| Zeravshan Basin Agency 771.000 ha | Samarkand, Bukhara and Navoi Provinces |
| Kashkadarja Basin Agency 504.000 ha | Kashkadarja Province |
| Surkhandarja Basin Agency 328.000 ha | Surkhandarja Province |
| Amu-Darja Basin Agency 776.000 ha | Khorezm Province and Karakalpakstan Autonomous Province |

It is important to point out how this promoted reform was challenging for the bureaucratic and state-oriented Uzbek institutions. In some cases, according to the reform's proposal, the management of the large-scale irrigation systems — controlled for several decades by the same province water department's members — would have had to be unified under a new authority, modifying the old institutional assets. This was the case in the Fergana and Zeravshan valleys, where it was necessary to unify three province administrations into a singlenew one, and in the case of Amu-Darja, where Khorezm was joined together with the autonomous province of Karakalpakstan. Therefore, this measure promoted by the Cabinet of Ministers at the national level inevitably led to misunderstandings and discontents among the province water departments' members. Decisions on water allocation and distribution have always been influenced by the water and agricultural departments; this measure would lead to a decrease in those departments' influence in decision-making processes. Accordingly, the provincial governors *(hokim)* jointly prepared a proposal which hampered the Program's enactment already designed by the Ministry of Agriculture and Water Resources, leading to significant changes in the reform's intent.[291]

---

[291] IWMI, 2012."cit.".

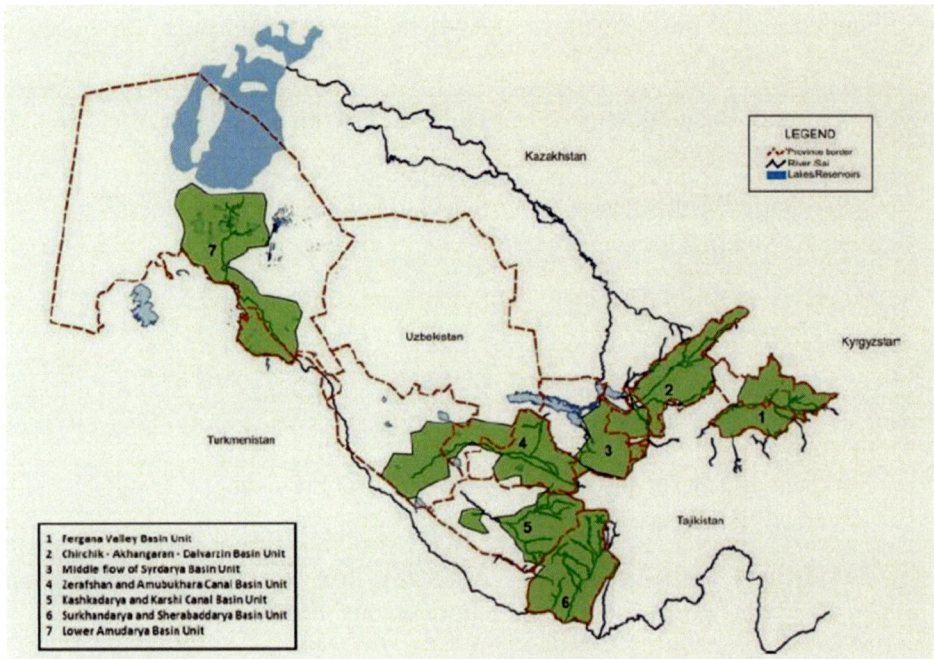

*FIG.10 : Thematic map of Uzbekistan showing the 2001's proposed BISAs and their boundaries; approx. scale:1:18.000.000, N↑(source: Wegerich, 2014).*

This proposal was featured by a conservative approach: the main aim of the province governors was to limit the changes concerning the organizational and territorial structures of the water authorities. Though no official papers and little data are available  regarding this idea enacted by the basin level actors, evidence has shown that the proposal was a move oriented towards avoiding the decrease of the decisional powers of province authorities in water resources management. In 2003, by the Decree n. 320 (21/07/2003), issued by the Cabinet of Ministries after the presidential one, the restructuring of the national water management according to basin principles was finally adopted: ten Irrigation Basin Management Authorities (BISA) were established instead of the seven initially proposed by the Minister in 2001.[292] The Cabinet of Ministers issued the following decree [293].

---

[292] G.W.P and UCC - Water, 2006."cit.".

[293] YALCIN, R., MOLLINGA, P., 2007. "cit.".

*"Accepting a proposal from The Ministry of Agriculture and Water Resources, Ministry of Economy and Ministry of Finance of Uzbekistan about creating the below-mentioned Irrigation Basin Management Authorities under the organizational structure of the water resources departments of the Ministry of Agriculture and Water Resources of Uzbekistan and its territorial subdivisions the following 11 main entities with their further subdivisions specified in the decree's appendixes shall be created"*

The Irrigation Basin Management Authorities (BISA) created by the Cabinet of Ministers' decree n.320, 2003 are as follows:

| |
|---|
| **Norin-Karadarja** Irrigation Basin Management Authority |
| **Norin-Syrdarja** Irrigation Basin Management Authority |
| **Syrdarja-Sokh** Irrigation Basin Management Authority |
| **Lower Syrdarja** Irrigation Basin Management Authority |
| **Chircik-Akangaran** Irrigation Basin Management Authority |
| **Amu-Surkhandarja** Irrigation Basin Management Authority |
| **Amu-Kashkadarja** Irrigation Basin Management Authority |
| **Amu-Bukhara** Irrigation Basin Management Authority |
| **Lower Amu-darja** Irrigation Basin Management Authority |
| **Zeravshan** Irrigation Basin Management Authority |
| **Main canal authority for Fergana Valley** with unified dispatch centre |

The newly created water structure abolished the 13 province water departments and the 163 district ones; these district authorities were replaced by 53 subdivided offices within the Irrigation Basin Management Authorities (BISA). Although this measure had been viewed by the international agencies as a real change towards water management according to *hydrographic* principles and political decentralization, analyzing the new structure the evidence has shown significant ambiguities. Looking at the territorial characteristics of the new BISAs' focusing specifically on the boundaries, it has emerged that five of the new authorities were created keeping the same boundaries of the previous province water departments; Norin-Karadarja, Norin-Syrdarja, Syrdarja-Sokh, Amu-

Kashkadarja and Amu-Surkhandarja BISAs were still based on administrative boundaries, hence without the changes required by the decree.[294] Furthermore, within the decree, the Zeravshan BISA — promoted in the 2001 program for the whole Zeravshan valley — has been divided into two different entities: Amu-Bukhara BISA for the lower valley, and Zeravshan BISA for the middle one. However, the outcome was that centralization reappeared, though in a different form; the new system is as centralized as previously, but with control now located in Tashkent. The newly established BISAs are directly responsible to the Ministry in Tashkent, with the difference that the basin authorities are no longer responsible to the local (province and district) governors (*hokim*); the BISAs' heads are now appointed directly by the Ministry of Agriculture and Water Resources instead of by the local government (*hokimyat*).[295] According to Yalzin and Mollinga (2007) this reform, oriented towards a hydrological-based administration, has been a measure of the Ministry of Agriculture and Water resources as a whole to reduce its dependency on province authorities' *(hokimyat)* influence. They claimed that the reform of the organizational structure has to be understood in the context of broader changes in the nature of the governance system to "depoliticise" certain sectors — that is, province and district *hokimyat* — to achieve efficient planning while maintaining centralized control.[296] Water resources management and allocation before 2003 was often hindered by interference from the province political authorities, creating specific differences and inequalities among the water users; hence the central authority had to break up the power of *hokim* to reduce the competition between the districts over water distribution. Therefore, the former Minister Djalalov was able to change the water organizational structure towards a fuzzy new national centralization, while ensuring the support of the international donors, and under the full approval of the ICWC, because of the promotion of some of the IWRM's principles. Although, as described above, each different BISA presents different territorial features, based still on province or hydrographic principles, also at the internal level, the Irrigation Systems Authorities (ISAs) were designed according to different features. Most of the ISAs were created based on the previous District Water Departments' (*Rayvodkhoz*) boundaries — how many and their area depending on the province's features. In other BISAs, specifically in those organ-

---

[294] Personal communication with IWMI, Tashkent, October 2012.

[295] Personal communication with GIZ, Tashkent, April 2012.

[296] YALCIN, R., MOLLINGA, P., 2007. "cit.".

ized on *hydrographic* principles, ISAs were designed *ex-novo* by the director and the water authority's administration. Some of the newly established ISAs territorially include the area for two districts, while others refer to the irrigation scheme's area, including different parts of the administrative units. Despite these apparently significant changes in water management structure, which occurred, as described above, in different ways depending on the regions and its province administration, the main tasks of the newly established authorities have not significantly changed compared to the former *Oblastvodkhoz* and *Rayvodkhoz*. Furthermore, in most cases the centralized, bureaucratic, and top-down approach was kept among the higher level authorities and the lower ones. Although, as claimed by Yalcin and Mollinga (2007) and other experts, the former Minister Djalalov, through the 2003 decree, had the willingness of a new centralization of basin water entities under the ministry's control, it must be underscored that the local governors (*hokim* )have been finally able to change the first reform proposal (2001), giving a  strong "local" impress to the BISAs' structural design, and partly adapting the basin agencies' territorial features to their will. Therefore, it is important to point out that, although the BISAs and ISAs tasks are nationally the same, the local administrations were able to partly or completely redesign their structure and boundaries. The Irrigation Basin Management Authorities annually receive a water quota (*limit*) for irrigation which has to be divided among the ISAs. Therefore, according to the decree, the BISAs' main tasks are to control and maintain the main canals and reservoirs through the Main Canal System Authority (MCS) and to divide and allocate the total amount of water to the ISAs, depending on the number of water users associations and their crops plans.[297] According to the agricultural systems' state quota, water allocation for cotton and wheat farming is privileged compared to the other crops. Several experts interviewed both at national and basin level claimed that the BISAs' staff and functions have not particularly changed if compared to the former province water departments; these tasks were carried out until 2003 by the district water departments, instead of by the ISAs.[298] Concerning the dismantling of the collective farms and cooperatives and the rise of peasant farmers and water users associations, the ISAs, in contrast  to the *rayvodkhoz,* are directly involved with water allocation to the WUAs. Irrigation systems authorities, in

---

[297] AMINOVA, M., ABDULLAEV, I., 2009. "cit.".

[298] Personal communication with GIZ, Tashkent, April 2012 and with Zeravshan BISA' staff, Samarkand, April 2012.

most cases newly designed by the BISAs' members, are responsible for the control and maintenance of the secondary canal network; the canals' supervision is divided among different working units to better control the hydro posts and the outlets. Annually, at the beginning of the cropping season, the ISAs stipulate a contract with the different WUAs, based on the water allocation quota; the WUA is subsequently responsible for water distribution depending on their crops plans. In contrast to other countries, such as neighbouring Kazakhstan and Kyrgyzstan, where the secondary canals' network control was handed, through leasing, from the district water departments to the WUAs, in Uzbekistan these facilities have been kept under state control. Therefore, the IMT process has been limited to the tertiary canals level, which is nowadays under the WUAs' administration. Although it was not officially mentioned in the 2003 decree on BISAs, according to the international donors' rationale supporting the IWRM, the newly set-up authorities, BISAs and ISAs, should have been accompanied by the creation of governance structures like the basin councils; these governance structures have been widely promoted by the Global Water Partnership and other organizations.[299] As both local experts and BISA and ISA members claimed, the basin councils have not yet been created in the newly established water authorities; hence, a governance structure does not exist to interface with the managerial level.[300] According to a Dargom ISA member (Zeravshan BISA), it is not necessary relevant for their authority to create a governance structure such as the basin or sub-basin councils. He claimed that they are able to discuss and address local issues through informal talks and meetings, hence, the set-up of councils is not necessary.[301]

### 4.2.2 Reshaping the IWRM rationale: the strong influence of government logics in the WUAs establishment

As mentioned at the beginning of this section, the second step of the "Program of measures on the improvement of irrigated lands for 2001–2010" discussed in 2001 focuses on the improvement of water management at the local level, and specifically on the establishment of the worldwide sponsored Water Users Asso-

---

[299] G.W.P and UCC-Water, 2006. *"Road Map" Planned Steps towards Realization of the Integrated Water Resources Management Principles and Rationale of the Essential Activities in the Republic of Uzbekistan.* Tashkent.

[300] Personal communication with GIZ, Tashkent, April 2012 and with Zeravshan BISA' staff, Samarkand, April 2012

[301] Personal communication with Dargom ISA members, Urgut (Samarkand province), October 2012.

ciations (WUAs), replacing the water tasks of the former collective farms. As early as the end of the 1990s, as stated by Wegerich (2000), the government of Uzbekistan tried to create pilot-project WUAs through the assistance of SANIIRI and the European Technical Assistance to the Commonwealth of Independent States (TACIS), even if without an official legal framework. Nevertheless, although with some evident lacks, twelve "informal" WUAs were established.[302] According to the transitional agricultural and water context, the WUAs were intended to improve water allocation, its equity and efficiency, and support the participation and integration of the water users in the decision-making processes; furthermore, their financial sustainability through the Irrigation Service Fee (ISF) could lead to the decrease of governmental expenditures. Although the WUAs' establishment was meant to promote bottom-up practices, coming from the water users, according to Yalcin and Mollinga (2007) in Uzbekistan the initiative for the WUAs' establishment did not come from the farmers but rather from the government. Farmers were asked to become members of the associations, and their leaders and technical staff were selected under close supervision of the local authorities or the province/district departments of the Ministry of Agriculture and Water Resources.[303] Therefore, despite the international donors' wisdom, the Uzbek model of establishing WUAs seemed to be an example of a top-down creation of a new organization at the local level. It reflected the authoritarian nature of the state in general and demonstrated that "reform" does not necessarily mean reduction of state control.[304] This process can be defined as a new form of controlling water allocation at the local level, after the former collective farms' dismantling, through members politically close to the local authorities. The first real WUA in Uzbekistan was established in Khorezm province in 2000 by a local initiative coming from the district water department and formalized by the Ministry (MAWR). As the agricultural transitional process was just beginning (the shift from *shirkat* to private farms was decreed by law in 1998), the idea was to transfer the irrigation responsibilities from the *shirkat* to the newly established WUAs based on the former administrative boundaries.[305] Thus, on January 5, 2002, the Cabinet of Ministers of Uzbekistan issued Decree n.8 to formalize the dismantling of the *shirkat* and the establishment of the

---

[302] WEGERICH, K., 2000. *Water Users Associations in Uzbekistan and Kirghizstan: Study on Conditions for Sustainable Development,* Occasional Paper n.32, SOAS, University of London.

[303] YALCIN, R., MOLLINGA, P., 2007. "cit.".

[304] YALCIN, R., MOLLINGA, P., 2007. "cit".

[305] YALCIN, R., MOLLINGA, P., 2007. "cit.".

WUAs countrywide; in absence of a law, the decree was the first legal measure allowing the WUAs' establishment.[306] However the decree did not clearly specify the work and the status of the associations, but just for which purposes they should operate. Due to the farmers and former collective-farms members' lack of experience, the government asked for the collaboration of international donors, such as the World Bank and the Asian Development Bank, to assist local authorities in the WUAs' establishment in several Uzbek provinces.[307] Water users associations were registered in justice departments as non-profit associations of water users based on administrative principles covering an area ranging on average from 1500 to 3000 ha and aimed at water allocation and maintenance at the tertiary canals level. Whereas during 2000–2002 the first created WUAs were considered experimental cases, by 2003 they were significantly strengthened, thanks to the aid of the development agencies. Since the issue of decree n.320 of 2003 on BISAs, the established WUAs stipulated contracts with local ISAs for annual water supply. Therefore the water allocation is left up to the main hydro technicians — part of the permanent staff that also includes the director and the accountant — and to the *miraab,* often hired from the WUAs from April to October. A challenging issue for the WUAs' fair performance is the management of the financial budgetthat should be gained from the farmers through the Irrigation Service Fee (ISF) which is intended to cover all the operation and maintenance expenditures. ISF started to be practiced in Uzbekistan in 2001, when the first experimental WUAs were created; however, according to Yakubov (2007), it still represents a great challenge because many farmers are not able or do not want to pay the water service fees.[308] He added that the fee collection rate in recent studies conducted in 2009 was on average 50%; this rate was also confirmed by Sehring (2006) in her research on WUAs in Tajikistan. The reasons for non-payment by the farmers range from inadequate and unequal water delivery service to lack of money due to unsold crops; furthermore, it is necessary to consider that the idea of free water for, based on free resources allocation from the state, is still widespread throughout Uzbekistan. With the aim of strengthening the local communities' role in the creation of new WUAs,at the end of 2003  the USAID and the WB  started financing a project named "Community Empowerment Network" which sponsored round-table discussions and

---

[306] IWMI, 2012."cit.".
[307] Personal communication with GIZ, Tashkent, April 2012.
[308] YAKUBOV, M., 2012. "cit.".

meetingswithin the organization; at the heart of the project was the idea of developing WUAs based on democratic principles and community empowerment with the hope of introducing a sort of "bottom-up revolution" at the local level to change the behaviour of local actors, such as farmers and irrigation-management staff.[309] The number of established WUAs throughout Uzbekistan increased from 86 in 2002 to 887 in 2005 and reached 1676 entities in 2008.[310] The rise of water users associations coincided with the dismantling process of the *shirkat* into private farms — on average ranging from 10 to 20 ha — which was almost completed in 2007. Nevertheless, during 2005–2006, the activities of some of the international donors — in particular those carried out by USAID — clashed with government policies based on strong control and influence on the WUAs' establishment process and were stopped. Despite this, other organizations, such as the Global Water Partnership (GWP), continued inducing the government to support the transition from a rigid administrative system of water management to a decentralized one, with massive public participation in water management. Moreover, GWP and others started supporting the establishment of the WUAs based on *hydrographic* principles to better manage water control and allocation. In 2009 a new measure was issued by the government to strengthen the WUAs' institutional framework; the Law (n. 240, 25/12/2009) "On introducing amendments and addenda to some legislative acts of Uzbekistan in connection with the deepening of economic reforms in water management" led to significant changes and amendments to the Law "On water and water use" issued in 1993. One of the adaptations is Article 18-2 which stipulates that "WUAs are created mainly by hydrographical principle or other conditions that ensure the sustainable management and use of water resources". Furthermore, according to the law, the founders of the WUAs may be farmers and/or plot owners as well as other water consumers.[311] As it was stated by the experts working in the international agencies, few WUAs throughout Uzbekistan have been established according to hydrological principles, hence, since the issue of the law, this condition might be strengthened in the near future. Another debated issue regards the WUAs' founders, who often are appointed or strongly influenced by the local authorities or were heads or members of the former district water departments.[312] In both of the enacted measures regarding WUAs (decree n.8-2002 and

---

[309] YALCIN, R., MOLLINGA, P., 2007. "cit.".

[310] GUNCHINMAA, T., YAKUBOV, M., 2009."cit.".

[311] IWMI, 2012."cit.".

[312] Personal communication with SDC and GIZ members, Tashkent, April 2012.

law n.240-2009) no references to governance structures characterized by a participatory approach were mentioned, even though it is one of the fundamental pillars of the IWRM, sponsored by the Global Water Partnership and others organizations. These issues will be widely debated in the next chapter, when focusing on data collected in the Middle Zeravshan valley.

### 4.2.3 Supporting the IWRM rationale: the donors based IWRM-Fergana valley and RESP II projects

According to a IWMI report (2012), even though since the 2000s several donors have supported the WUAs'establishment and strengthened it through development projects based on the IMT and the IWRM framework, nowadays only a few of them are still operating within Uzbekistan.[313] The most important project designed in the last decade and based on the IWRM's framework, support, and implementation is the IWRM-Fergana Project. This project has been working since 2003 and also involves some neighbouring regions in Tajikistan and Kyrgyzstan; it has been promoted by the Scientific Information Centre of the ICWC and the International Water Management Institute (IWMI), and mainly funded by the Swiss Agency for International Development and Cooperation (SDC).[314] Due to the importance of the Fergana Valley for Central Asian transboundary stability and water management, the project started promoting the IWRM pillars and strengthening the established organizations from basin to local levels, such as the basin agencies and the WUAs. The project aims to involve social groups and stakeholders in water processes through the concepts of integration and participation: water users groups (WUGs) were supported in the project area, involving the small farmers and household plot owners at the tertiary canal level. In addition, in recent years the establishment of WUAs based on hydrographic principles and characterized by participatory governance structures were strongly supported. At the basin level, the main canals — South Fergana, Aravan-Akbura canal and Khodja-Bakirgan — are controlled by BISAs and by a new organizational structure named Canal Administrations (CA), which works in collaboration with the Canal Water Users Union (CWUU), including all the water users supplied by the same canal. These institutional innovations generally improved water management, reducing potential disputes and facilitating water allocation. For a decade, these topics have been deeply analysed

---

[313] IWMI, 2012. "cit.".
[314] DUKHOVNY, V. et al. 2009. "cit.".

within the IWRM-Fergana Valley project, by several scholars, such as Wegerich, Yakubov, and others.[315] Another international project has been designed throughout Uzbekistan since 2009, named Rural Enterprise Support Project II (RESP II), funded by the WB and the SDC and implemented by SIC/ICWC. The project development objective is to increase the productivity and the financial and environmental sustainability of agriculture as well as the profitability of agribusiness in the project area. Objectives will be achieved through the provision of financial and capacity-building support to the farmers and agribusinesses in seven provinces as well as improve irrigation-service delivery through rehabilitation of Irrigation & Drainage infrastructures and the strengthening of WUAs based on *hydrographic* principles in seven districts within seven different provinces of Uzbekistan.[316] One of the selected districts is Pastdargom, lying in Samarkand province and chosen as a field-research area. In the last years, training programs and seminars for the water users were conducted, part of the deteriorated irrigation schemes were restored, and 62 new WUAs were established within the seven districts.[317] According to IWMI, which collected data from the MAWR, in 2012, 1487 WUAs were operating, covering a total irrigated area of 3.377.900 ha.

---

[315] DUKHOVNY, V. et al., 2009. "cit.".

WEGERICH et al, 2012. Is It Possible to Shift to Hydrological Boundaries? The Fergana Valley Meshed System, *International Journal of Water Resource Development,* vol.8, n.3.

YAKUBOV, M., UL HASSAN, M., 2007. Mainstream Rural Poor in Water Resource Management: Preliminary Lessons of a Bottom-Up Development Approach in Central Asia, *Irrigation and Drainage,* 56, 261-276. YAKUBOV, M., 2012."cit.".

GUNCHINMAA, T., YAKUBOV, M., 2009. "cit".

BICHSEL, C., 2009. *Conflicts Transformation in Central Asia: Irrigation Disputes in the Fergana Valley,* Routledge, London & New York.

[316] WORLD BANK, 2012. *Proposed Project Restructuring of Second Rural Enterprise Support Project,* Restructuring Paper from the WB to the Republic of Uzbekistan.

[317] WORLD BANK, 2011. *Implementation Status and Results, Uzbekistan, Rural Enterprise Support Project II,* Report n. ISR6177.

## 4.3 THE WATER REFORMS PROCESS AT THE BASIN / LOCAL LEVEL IN KAZAKHSTAN

### 4.3.1 A conflicting relation between the IWRM rationale and the local logics at the basin level

As emerged in Uzbekistan, since the mid-1990s, in Kazakhstan the international donors have started supporting, in collaboration with the government, the IMT process and the WUAs establishment, a transition path oriented towards adopting the IWRM framework. As it was analysed in the previous chapter, the dismantling of former collective farms and the rise of peasant farms have occurred in a shorter time compared to Uzbekistan; the Land Code was enacted in 1995 and after the Bankruptcy Law was issued (1998), the agricultural sector's transition process was almost completed. Therefore, already by the end of the 1990s it was necessary to reform the existing water management structure, as it was unable to deal with the thousands of private farmers, through the formalization of the IMT and the subsequent WUAs establishment.[318] Although the IMT process in Kazakhstan began 15 years ago, nowadays at the basin/local level an ambiguous and debated water management structure emerges. At the basin level, two main organizations are involved in water resources monitoring and control: the River Basin Agencies (*BWO)* and the Republican State Enterprises (*RGP*). Although, as Zimina (2002) states, according to the 1993 Water Code, River Basin Agencies are the primary water management agencies in Kazakhstan; in 1996 a Decree on the Differentiation of Functions between the two agencies was enacted in order to clarify the respective tasks and prevent disputes.[319] As it was mentioned in the previous chapter, the *BWO* are the eight branches of the Committee of Water Resources (under the Ministry of Agriculture) covering the whole Kazakhstan territory, organized according to basin principles. For instance, Balkash-Alakol River basin includes the Almaty province and parts of Karaganda and East Kazakhstan province, and Aral-Syrdarja *BWO* covers South-Kazakhstan and Kizylorda provinces' territory. These organizations receive the annual total water amount from the Committee of Water Resources and are responsible for monitoring water resources and consumption according to basin principles, overseeing water quality and pollution levels and controlling inter-province and inter-state water reservoirs. Although their role ought to have

---

[318] WEGERICH, K., 2008. "cit.".

   BURGER, M., 1998. "cit.".

[319] ZIMINA, L., 2002. "cit.".

been strengthened by the 2003 Water Code, supporting basin principles organizations and the IWRM, interviewed members claimed that in the last ten years the agencies have not been significantly reinforced.[320] In contrast, the Republican State Enterprises were formalized with their current status through the 1996 Decree; until the mid-1990s these organizations were funded by the province financial budget and named Province Water Departments (*Oblastvodkhoz),* while after the measure by the governmental budget. These organizations are therefore based on administrative principles (province territories) and are responsible for operation and maintenance of the primary level water system units (main canals and reservoirs), improvement of technical infrastructures conditions, and water allocation to the local level water organizations. Although the organizations' responsibilities have been formalized, Zimina (2002) claims that the 1996 Decree reinforced the Republican State Enterprise in respect to the River Basin Agencies; in the years that followed the decree, River Basin Agencies have been underfunded and thus unable to fulfil their responsibilities. Zimina added that a paradoxical situation emerged because the technical body possessed more human and financial capacities than the controlling ones.[321] These institutional and organizational conditions contrasted both with the 1993 Water code and the governmental will to support the water management based on basin principles and the IWRM framework. Since 2004, as partially analysed in the previous chapter, the international project "National IWRM and Efficiency Plan for Kazakhstan" sponsored by the UNDP and the GWP, aimed to support water management according to basin principles, therefore strengthening the River Basin Agencies and establishing the River Basin Councils (RBCs) for each agency.[322] As mentioned by the members of the Aral Syrdarja Basin Agency, on the one hand, the organizations were reinforced, in particular according to an institutional/political perspective; on the other hand, their tasks and the organizational features have not significantly improved in the last years.[323] The other objective of the project, with the aim of setting up a governance structure within the River Basin Agencies, has been the establishment of the River Basin Councils. These councils should increase and promote the participation of all the water users in decision-making processes and water use, involving members from the province and district governments (*Akimyat),* from

---

[320] Personal communication with Aral-SyrDarja River Basin Agency' members, Shymkent, April 2012.

[321] ZIMINA, L.,2003. "cit.".

[322] UNDP, Government of the Republic of Kazakhstan, 2004. "cit.".

[323] Personal communication with Aral-SyrDarja River Basin Agency' members, Shymkent, April 2012.

128

basin level organizations, such as RGP and BWO, and from local ones, such as District water departments and WUAs members as well as farmers and household plot owners.[324] The council's organization officially started in 2005 with training sessions and meetings between the donors, the Committee of Water Resources, and members of the water organizations. The first RBC was established in 2006 in Balkash-Alakol River Basin Agency. This establishment process continued with the "National IWRM and Efficiency Plan for Kazakhstan" project extension, initiated in 2008 and divided into two phases, 2008–2010 and 2010–2025. Nevertheless, although the donors actions regarding the empowerment of *BWO* and the establishment of RBC have been developing for four years, in 2008 they still confirmed an organizational immaturity and a sectoral fragmentation in the reform's implementation. Furthermore both the Committee of Water Resources and the River Basin Agencies did not fully understand the need of the councils and were quite reluctant to interact with non-governmental stakeholders.[325] Finally, as stated by the Hydro-melioration State Enterprise members, the eight River Basin Councils were established at the end of 2008, though with weak institutional and organizational structures. No elections among the members were organized to decide the RBC head, which is the self-appointed director of the Basin Agency; on average those meetings are organized twice a year.[326] In some basin units, as claimed by the Aral-Syrdarja Basin Agency members, sub-basin councils were organized according to the specific irrigation system's territory in order to more fully involve the water users, as occurred in both Balkash-Alakol and Aral-Syrdarja Basin Agencies.[327] Nevertheless from the interviews conducted with the water users in the villages, lack of interest emerged regarding the river basin councils; most of them are not aware of those organizations, included the WUAs directors, and furthermore, some of them stated that they were not involved in the councils of either the development agencies or River Basin Agencies. In addition some of the water users claimed that those councils are not necessary for the improvement of water management practices and it would be more relevant to strengthen the basin/province level

---

[324] UNDP, Goverment of the Republic of Kazakhstan, 2004. "cit.".
Personal communication with Hydro-melioration State Enterprise members, Shymkent, November 2012.
[325] UNDP, 2007."cit.".
[326] Personal communication with Hydro-melioration State Enterprise members, Shymkent, November 2012.
[327] Personal communication with Aral-SyrDarja River Basin Agency' members, Shymkent, April 2012.

departments which deal with irrigation systems' operation and maintenance.[328] In the last years, the River Basin Agencies' members claim that their organizational framework has not improved, nor their financial availability, even though the ongoing international project; in contrast the Republican State Enterprise's members have claimed different opinions. According to the interviewed members of the South-Kazakhstan agency (*RGP-Iujvodkhoz*), in the last few years the government funds significantly increased, due to the following reasons: firstly, due to a general improvement of the Kazakh national economy and the subsequent government concern to invest in the state water departments and infrastructures; and secondly, due to the cessation of government funding to the district water departments, which occurred at different times some years ago. Therefore the Republican State Enterprises were able to restore some of the primary-level water systems and improve their tasks in water allocation at the local level.[329]

### 4.3.2 The two water management rationales at the local level: the WUAs and the district water departments

Just as the water management's structure at basin level is characterized by two bodies, formalized according to the Decree "On the differentiation of functions", at the district/local level two organizations with similar tasks deal with water management and allocation. Although, as previously emphasized, the IMT process has been supported in Kazakhstan since the mid-1990s, with the main aim of formalizing the WUAs establishment at the local level, the evidence has shown that the process is not yet complete; district water departments and state entities are still operating, structuring the water context among governmental and non-governmental bodies. The WUAs' establishment processes in Kazakhstan have been analysed and debated by several scholars regarding procedures and times. As early as 1995, the Harvard Institute for International Development (HIID), with the financial support of USAID, began assisting water officials and farmers to improve water resources management and irrigation. That same year, these organizations in collaboration with the Committee of Water Resources launched the project "Improved Management of Water Resources" to support and facilitate the IMT process. The district water departments, dealing with water man-

---

[328] Personal communication with the water users (farmers, WUAs member and staff), Southkazakhstan province, April-May / November-December 2012.

[329] Personal communication with Republican State Enterprise' members, *Iujvodkhoz*, Shymkent, November 2012.

agement and allocation at the district level since the Soviet Union, lacked both the funds and the expertise needed to manage and maintain the irrigation systems. Furthermore, they were designed to deal with former collective farms and, therefore, those departments were no longer able to handle the rise of peasant farmers. According to Burger, HIID member, in 1996 the Kazakh government launched a program of public tenders to offer the associations of water users the right to bid on water facilities management; no significant results emerged from this measure because of economic-political issues, such as the private stakeholders' inability to get external funds to upkeep such facilities.[330] In order to strengthen the challenging transition processes, in 1997 a guidebook including all the recommendations regarding the WUAs' establishment was provided by the HIID/USAID project to the farmers trying to cope with water management's lacks and to help the water users in an institutional and organizational perspective to create a non-governmental association. Different opinions have emerged among scholars about when the first WUAs began working in Kazakhstan; according to Wegerich (2008), there were misinterpretations about the formal and informal status of the WUAs and about who could be a member of the association.[331] While Zimina in 2002 argued that the first WUA in Kazakhstan was already established in 1993, as a water user association without a formal organization, Mott Macdonald's report states that the WUAs started working in part of Kazakhstan in 1996.[332] Nevertheless it should be underlined that the WUAs' establishment in those years, 1996–1998, coincided with the set-up of a development project at the local level by the international donors — World Bank, Asian Development Bank, and USAID — supporting the formalization of IMT and water facilities' restructuring. Credits and financial aid were provided to the farmers, often the members of the former collective farms, to establish a WUA with a stable organizational structure. Furthermore, in accordance with socio-environment sustainability, strongly promoted both by the IWRM framework and the IMT, in 1997 water fees for agricultural water users were introduced, calculating the water use by the cubic metre. Nevertheless, as the irrigation schemes were built during the Soviet Union for large collective farms, often water consumption has been just estimated rather than measured, leading to inequi-

---

[330] BURGER, R., 1998. "cit.".

[331] WEGERICH, K., 2008. "cit.".

[332] ZIMINA, L., 2003. "cit.".
   MOTTMACDONALD-DFID, 2003.*Privatization / Transfer of Irrigation Management in Central Asia,* final report .

ties and abuses among the farmers.[333] The weakness of the newly established WUAs as far as their performance of technical/organizational tasksshould also be considered in terms of their ability to provide water allocation to the users according to schedules. In addition, as Mott Macdonald's report partly claims, a WUA is not able to workfairly without an appropriate legal framework which institutionalises its status and responsibilities, such as management and governance.[334] Therefore, recognizing the inadequacy of the 1993 Water Code to strengthen the reforms oriented to the IMT, in July 2003 the government of Kazakhstan, with the support of the UNDP, enacted a new Water Code; within this measure, as previously pointed out, the IWRM framework has been formalized as the strategic and challenging pattern for future development in the water sector oriented towards socio and environmental sustainability. Following this perspective, the same year, Law n. 404-II issued by the Cabinet of Ministries of Kazakhstan formalized the WUAs as Rural Consumers Cooperatives of Water Users (*SPKV*), providing an official status to the existing associations and giving the farmers a simpler route for self-organizing a WUA.[335] According to the enacted law, the WUAs have to register as non-profit organization in the district judgement department specifying the director, members, and its features. Regarding their territory, despite that ones based on *hydrographic* principles officially were supported according to the IWRM, the Kazakh WUAs were organized referring both to administrative boundaries — such as those established for the former collective farms and districts — and *hydrographic* ones, covering on average 1500–2500 ha. In most cases the established WUAs refer to the old state or collective farms' territories. As highlighted in the enacted law, the WUAs should be responsible for their operation and maintenance (O&M), hence, water allocation to the farmers and maintenance of the irrigation facilities at the tertiary and secondary levels, through water fee collection from the members. Regarding the water charges, the rates are fixed by the State Anti-monopoly Committee and subsequently the WUAs add a small amount to cover service charges; on average the water fee in 2012 ranged from 220 to 350 tenge (1.1 to 1.8 USD) for 1000 cubic meter, depending on the water source (river, canals, or reservoirs) and on the organization's features. Whereas the tertiary canals are managed by the WUAs without any kind of official permission, the secondary canals' sys-

[333] WEGERICH, K., 2008. "cit.".

[334] MOTTMACDONAL-DFID, 2003. "cit.".

[335] MOLLE, F., 2007. "cit.".

tem is leased from the district water departments through a contract which allows its operation and maintenance for a long-term period (5 to 10 years). According to Salman (1997), the autonomous irrigation agencies, such as the WUAs, should provide better services to the users, promoting governance and providing a fair share of water in a timely manner.[336] It is important to point out that non-governmental organizations such as the WUAs are totally new in the Kazakh socio-cultural environment, where for decades the farmers were assisted by state organizations; therefore, on the one hand, the human/social cohesion inherited from collective farms could represent an advantage in establishing a WUA, on the other hand, the water users have to learn to work independently, concerning both technical and organizational issues, and to participate in the decision-making processes. Farmers are normally not used to the internationally supported participatory approaches like organizing councils and debating issues according to democratic principles. Wegerich claims that the 2003 Law regarding the WUAs left unclear how the governance structure should operate and questions about its separation from the management structure.[337] Whereas at the basin level the governance-participatory approach has been strongly supported through the establishment of the River Basin Councils, focusing on the WUAs, these principles have not been officially mentioned; solely in some international development projects designed by the WB, ADB or USAID — for instance in Makhtaral district (South-Kazakhstan province) — the participatory approach has been sponsored in the established WUAs. Although hundreds of associations have been designed and formalized since 2003, according to the development projects and the farmers' organizational and financial availability, at the district level the district water departments continued operating, providing water to private farmers or cooperatives not involved in the WUAs. Focusing on the Makhtaral and Otrar districts, both in South-Kazakhstan province, evidence has shown how the action of the international donors, working in Makhtaral since the 1997, facilitated the establishment of the WUAs and thestrengthening of the IMT, leading to a decrease of the district water department's role. The context significantly differs in Otrar district where no projects were designed and the creation of WUAs seemed more difficult and challenging. Nevertheless, it is important to underline the challenging process that emerged since the WUAs' establishment and the effectiveness of theirsubsequent performance. Both

---

[336] SALMAN, R., 2007. "cit.".
[337] WEGERICH, K., 2008. "cit.".

Zimina (2002) and Wegerich (2008) claimed that several WUAs in Kazakhstan did not operate fairly, and some of them failed, due to the strong political influences of the local state actors in their organizational structure as well as lack of bottom support from the farmers. Moreover, the IMT have been ill-planned and the withdrawal of the state too rapid, without considering local realities.[338] What emerged from data collection and interviews in the last years is that the WUAs have not been assisted in their organizational development, neither from the international donors nor the government; whereas ten years ago, the WUAs were supported by different projects and financial aid programs, nowadays most of those activities have stopped working. Therefore, the WUAs, which rely on technical knowledge and financial availability, often directed by heads strictly connected with local state power, were able to operatefairly, while other associations that do not hold certain requirements stopped working or failed. For instance, in those conditions where the WUAs were not able to provide water in time or to allocate the scheduled water amount, interviewed water users generally claimed a loss of confidence in WUAs, in particular concerning more expensive water charges, if compared to the state departments; furthermore, they added that in those conditions it is ultimately better to deal with the district water departments.[339] As previously mentioned, despite the 2003 Law on WUAs, the district water departments have been performing the same tasks in the last decade as they have since the 1990s: operation and maintenance of the secondary canals level and water allocation to the farmers not involved in the WUAs. The water users who were members of a dismantled WUA, rescinded on the hired secondary canals' network contract with the district judgment court and resigned the annual water agreement with the district water departments. Therefore, according to the interviews conducted with the departments' members, some disputes regarding power and rights concerning water management and allocation emerged among them and the WUAs' members; furthermore, some of them have not expressed any kind of support for the IMT and do not regret the WUAs dismantling.[340] Hence the evidence highlighted how some of the state authorities at the district level have not effectively supported the IMT process heavily sponsored by the government at the national level. In the districts where

---

[338] ZIMINA, L., 2003. "cit.".

WEGERICH, K., 2008. "cit.".

[339] Personal communication with the water users, South-Kazakhstan and Almaty provinces, 2012.

[340] Personal communication with district water departments' members, South-Kazakhstan province, 2012.

several WUAs were dismantled, the district water departments, despite financial shortages, regained a powerful role in water management and allocation at the local level supporting the "top-down" approach, which for a decade the international donors have tried to replace with "bottom-up" practices. Nevertheless, as it was claimed by the Republican State Enterprises and by some of the interviewed directors of the district water departments, those organizations will be soon dismantled, as already occurred in some districts — for instance, in Tyulkibas (South-Kazakhstan province), which is one of the administrative entities which have been selected for fieldwork.[341] This reorganization of the water management structure at the district level, sponsored by the government, is due to the lack of an adequate financial budget for the district authorities who are no longer able to cover the expenditures for operation and maintenance. At present, it is not certain when this reorganization process will be completed, but probably it will be, as claimed by the interviewed experts, in 2014 or 2015.[342] It will lead to substantial changes in water management at the district level which will be analysed in depth in the next chapter, focusing on the selected districts.

---

[341] Personal communication with district water departments and Republican State Enterprises' members, South-Kazakhstan province, 2012.

[342] Personal communication with Republican State Enterprises and BWO Aral-Syrdarja' members, 2012.

# 5. THE LOGICS OF BASIN / LOCAL LEVEL WATER REFORMS IN UZBEKISTAN: THE MIDDLE ZERAVSHAN VALLEY

*FIG. 11: Satellite image (Landsat 5) representing the Middle Zeravshan valley (source: Uzbek-Italian Archeological project: Samarkand and its territory).*

## 5.1 THE GEOGRAPHICAL OVERVIEW: AN HYDRAULIC TERRITORY

### 5.1.1 The river and the physical context of the valley

As mentioned in the previous chapter, the middle Zeravshan valley was selected as the Uzbek case study for multiple reasons, ranging from territorial to economic-political ones; this section of the valley which lies in central-eastern Uzbekistan, is one of the largest and most important irrigated areas of the country and of the whole Central Asian region since ancient times. Its territorial and political development dates back to the second century BC when the Sogd civilization settled in those lands.[343] Zeravshan river is the third longest and most important river of the Central Asian region and it flows in a West-East direction at

---

[343] ISAMIDDINOV, M., 2002. *The Irrigation Development of the Samarkandian Sogd in the Ancient Times*, Izadeltstvo, Tashkent.

136

a latitude ranging from 39° to 40° N. The catchment of the river has an area of, on average, 143.000 kmq and it is divided into two parts: the upper narrow river valley in Tajikistan and the middle and lower basin plains in Uzbekistan; the Uzbek part of the catchment alone covers 131,000 kmq (90% of the entire basin).[344] Zeravshan river originates in Hissar/Zeravshan mountains, altitude 2750 meters a.s.l., with the first 300 km flowing in northern Tajikistan and ending after a total length of 740 km in Bukhara province; the average annual run-off is 5.3 km3 (FIG. 12).[345]

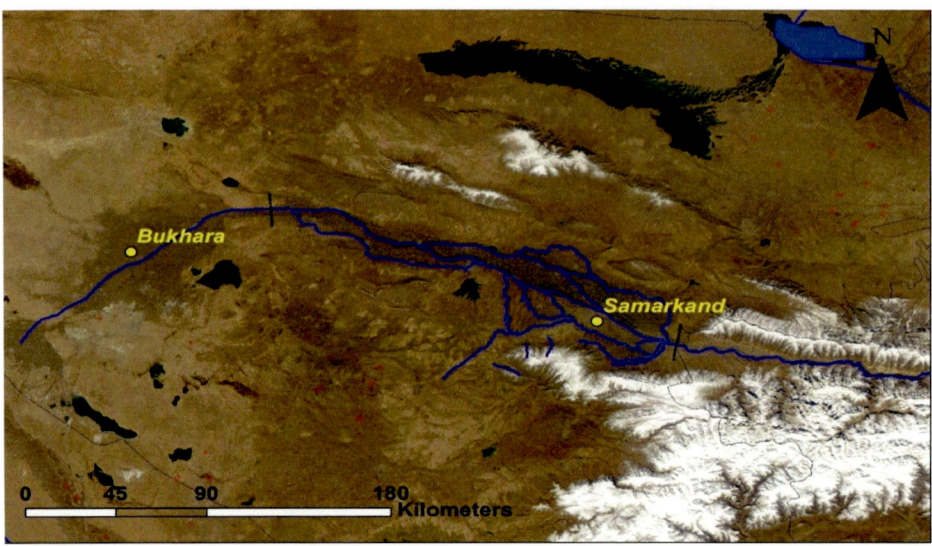

*FIG. 12: GIS elaboration of a satellite image (NASA-Modis, 2003) representing the three river sections, upstream, middle and downstream (separated by a black vertical line); nowadays the river (blue line) ends before reaching Bukhara, (approx. scale 1:4.000.000, N↑).*

The river, due to its physical-climatic conditions, is characterized by a simple regime with two hydrological seasons: a flood season divided into snow melt in spring and ice melt in summer, which is the highest, and the low water period corresponding towintertime. In the past, the river was a tributary of the Amu-Darja, but for several decades, due to the massive water use for irrigated agricul-

---

[344] FEDCHENKO, F., 1870. Topographical Sketch of the Zeravshan Valley, *Journal of the Royal Geographical Society of London,* 40.

[345] OLSSON, O., et al., 2010. Identification of the effective water availability from stream-flows in the Zeravshan river basin, Central Asia, *Journal of Hydrology,* 390, 170-177.

ture in its middle course, it never reaches Bukhara province. The Zeravshan river course can be divided into three sections: the upstream lying entirely in Tajikistan, the middle from the border until Navoi city, and the lower one until the flows' end which varies depending on the season and related runoff and water use.The Middle Zeravshan valley is administratively mostly included in the Samarkand province and the valley sectioncan be considered as lying from the 1$^{st}$ May Dam in the east (on the border between Tajikistan and Uzbekistan) to Navoi in the west. This territory includes most of the Central Asian physical-environmental features, such as irrigated plains, steppes, foothill areas, and mountains. The irrigated area nowadays stretches approximately 50–60 km N-S and 200 km E-W and it is surrounded in the eastern and northern part by the Zeravshan and Turkestan Ranges respectively, and in the western-southern part by the steppes which separate the Zeravshan valley from the Kashkadarja irrigated area. The eastern part of the valley is featured by an altitude slope (North-South) from 1000 m. in the foothill areas to 700 m. above sea level on the riverbanks; the Zeravshan, due to slight slope E-W and to the soil's characteristics, can be classified as a braided river surrounded on its branches by the tugai forest, a typical endhemic species of the Central Asian region.[346]

*FIG. 13: 3D Elaboration of satellite image (Landsat 5), 2009, representing the Middle Zeravshan Valley observed in W-E direction (approx. Scale: 1:2.000.000); (courtesy Dr. B. Rondelli, Uzbek-Italian Archeological Project ).*

---

[346] BENSIDOUN, S., 1979. *Samarkand et la Vallé du Zeravchan,* Anthropos, Paris.

138

Focusing on the local annual rainfall, relevant for irrigation practices, it differs significantly according to the proximity to the mountains; it ranges approximately from 200 mm/y in the southern-western part of the middle valley, 330 mm/y in Samarkand city, to 400 mm/a in the eastern foothill areas.[347]

### 5.1.2 The hydraulic territory: the complex canals' network

The Middle Zeravshan valley, focusing on its territorial features, can be considered as a hydraulic territory because of the strong human actions towards natural resources that have been carried out since ancient times until the collapse of the Soviet Union, significantly increasing the total irrigated area.[348] Stride et al. (2009) adds that the valley is formed by a complex network of natural and artificial water courses crossing the plain and creating a series of *jazireh*, each one of which is characterized by different physical-ecological features (FIG.13).[349]

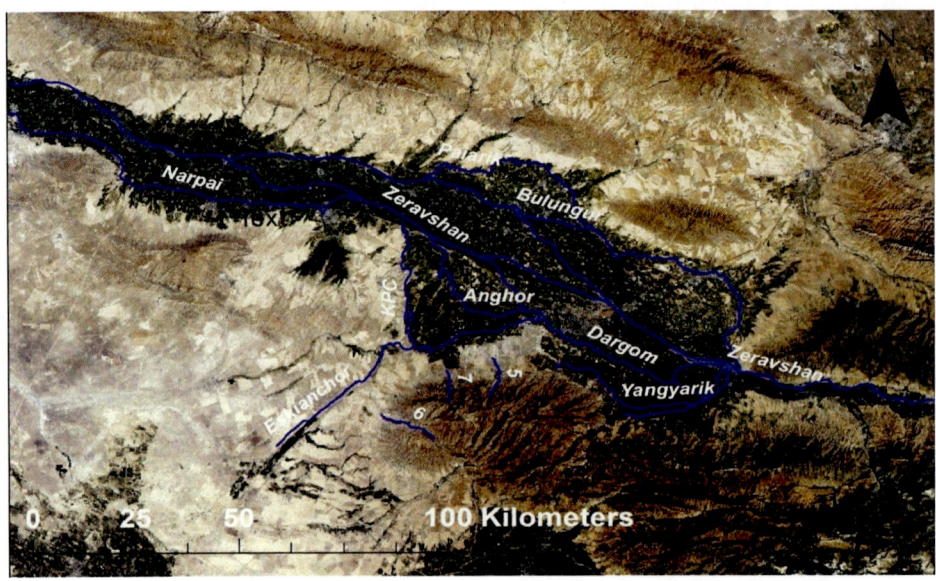

*FIG. 14: GIS elaboration of a satellite image (Landsat 5) representing the canals' system (blue lines), arising from the 1st May dam, of the Middle Zeravshan valley.*

---

[347] Data collected in Samarkand Meteorological Centre, Samarkand, May 2011.

[348] BETHEMONT, J., 1999. *Les Grandes Fleuves. Entre Nature et Société,* Armand Colin, Paris.

[349] STRIDE, S., et Al., 2009. Canals versus Horses: Political Power in the oasis of Samarkand, *World Archeology.* 41 (1).

Focusing on the Middle Zeravshan valley's canal network, this irrigation system arises today from the 1st May Dam, built in 1913 during the Tsarist Empire and located close to the Uzbek/Tajik border, stretching on both sides of the valley, wider in the southern part. Dargom and Bulungur are the most ancient, dating back to the Sogd civilization; Dargom canal, 100 km in length, arises from the Zeravshan, flowing on the left-southern branch of the valley and converging again in the river. According to both archaeologists and geographers, Dargom was created using an ancient riverbed of Zeravshan and the small riverbeds of the streams' (say), originating from the Chakylayan/Karatyube mountain slopes (Zeravshan range) and flowing down into the ancient Zeravshan riverbed.[350] On the northern bank, Bulungur canal was also built in ancient times; this water course originates from the 1st May Dam and flows for 90 km alongside the Turkestan range foothill area, converging in Zeravshan river (Karadarja branch). The river, close to Samarkand, naturally separates into two branches, Akdarja flowing south and Karadarja flowing north, creating an "island" named Miankal. The construction of Dargom and Bulungur canals enabled the rise and wide-spread use of irrigated agriculture in the valley, significantly increasing the total irrigated lands. Other canals which contributed to the extension of the irrigated plain during the centuries are the Paiarik, connected with Bulungur and Tuy-tartar on the northern side, and Yangiarik, Anghor, Narpai, and Eski-Anghor on the southern side (FIG.14). Whereas, on the one hand, the Narpai canal building enabled the extension of irrigated land in the lower part of the middle Zeravshan valley, the Yangiarik lead to the permanent Zeravshan water flow to the south-eastern plain and related Chakylayan foothills area. Yangiarik does not reach the Karatyube canal, and it is possible to observe in the satellite image that this area, not irrigated, is not feasible for agriculture. Since Zeravshan river is one of the most important water courses of the region, two canals, Eski-Anghor and Tuy-tartar, were designed to transfer water from the Zeravshan valleys to the neighboring ones which often suffer from water shortages: Eski-Anghor canal, lying in the southern branch and connected to the Kashkadarja irrigation system and Tuy-tartar canal, connected to the Jizzakh irrigation system in the north. The whole canals system was restored and extended during the Soviet Union and in particular beginning in the 1960s when, as mentioned in the third chapter, the

---

[350] BENSIDOUN, S., 1979. "cit.".
    ISAMIDDINOV, M., 2002. "cit.".
    STRIDE, S. et Al., 2009. "cit.".

Soviet government started carrying out the *hydraulic mission,* with the aim of expanding cotton-farming. In the southern part of the Zeravshan Valley three new canals were designed: KPC connected to Eski-Anghor, Obvodnoi Dargom connected to old Dargom, and Mashini canal, with the aim of lifting water to the Chakylayan's foothills area. In addition, sections of the other canals were lined, the outlets and main gates restored, and a new secondary water course was designed. Besides the canals, the Soviet government focused on territorial reorganization for the construction of huge pumping stations to lift water and increase the irrigated lands; the Mashini canal, mentioned above, is featured by five small pumps which lift water thirty meters higher from the Yangiarik to the canal which ultimately flows for 35 km. In the middle Zeravshan valley, 22 pumping stations were built since the 1960s; one of the most important pumping station lies in Urgut district, in the south-eastern side of the valley, named Urgut II.[351] This water infrastructure was built during the 1980s and through four pumping systems and two tubes lifts water from Obvodnoi Dargom 100 metres higher, where it then flows down into a secondary and tertiary canals level system. Urgut II pumping systems nowadays irrigates 2000 ha of land mostly oriented to wheat farming and fruit. The other systems are located in the central-southern part of the valley on the Eski-Anghor canal and in the western canal on the Narpai. According to the interviewed workers and experts at the national level, the pumping stations are managed by the Ministry of Energy through province and district departments, but are financially funded by the Ministry of Agriculture and Water Resources. As occurred in the whole Uzbekistan since the USSR's collapse, no huge hydraulic infrastructures have been built mostly due to lack of financial and technical resources; therefore, the total irrigated area has not increased in the last two decades, rather in some areas decreased due to the deterioration of water facilities. Whereas during the USSR era almost the whole irrigated area was oriented to cotton farming and fruit, since independence and the introduction of wheat farming, today the surface is approximately divided between the two main crops, still affected by state quotas, and tobacco and grapes. Nowadays according to the Zeravshan BISA's members, the Middle Zeravshan valley's irrigation system arising from the 1st May Dam reaches 547,000 ha.

---

[351] Personal communication with Zeravshan BISA' members and Urgut II workers, Urgut district, April 2011.

## 5.2 IMPLEMENTING THE RIVER BASIN UNIT: THE ZERAVSHAN IRRIGATION BASIN AGENCY

### 5.2.1 The local governmental logics in conflict with the IWRM rationale

As analysed in depth in the previous chapter, the reforms package enacted at the beginning of the 2000s by the Cabinet of Ministers oriented towards the IMT, and generally to the IWRM framework, strongly affected the water management reforms at the basin level; specifically at this level, the focus was on the creation of water bodies based on hydrographic principles. According to the "Program of measures on the improvement of irrigated lands for 2001–2010", designed in 2001, the province water department (*Oblastvodkhoz)* inherited from the Soviet Union and based on administrative principles, had to be reorganized according to the hydrographic principles. Therefore, according to the reforms' program, one Irrigation Basin Agency (Zeravshan BISA) should have been created for the whole Zeravshan valley's irrigated area as well as the Amu-Bukhara pumping systems (totalling 771.000 ha) including Samarkand, Navoi ,and Bukhara provinces and hence their water departments. Therefore, these authorities should have been dismantled leading to the establishment of a new administrative and management structure, supported and sponsored by the Ministry in Tashkent. As previously mentioned, as Yalcin and Mollinga (2007) claim, this measure was a move by the Ministry of Agriculture and Water Resources to reduce the power of province/district governments in water management and their institutional and organizational framework.[352] This measure inevitably led to political problems and disputes among the local governors which have managed province water control and allocation since the Soviet Union. Due to these controversial issues, in Zeravshan valley, as in several Uzbek regions, the 2001 reforms program's plan was disputed and finally modified according to the province governments' decisions. Instead of the first originally proposed plan to incorporate the three provinces — and therefore the Zeravshan basin as well as the Amu-Bukhara pumping-canal system — within one basin authority, two separate BISAs were established: Amu-Bukhara BISA and Zeravshan BISA. Those two bodies and all the other Uzbek BISAs were officialised after a couple of years within the previously debated Decree n.320 (21/7/2003). Ultimately, the established Zeravshan BISA territorially differs both from the 2001 idea and the old Samarkand province water department. The Zeravshan BISA includes the whole

---

[352] YALCIN, R., MOLLINGA, P., 2007. "cit.".

Samarkand province (67% of total catchment), four districts of Navoi (17%), three of Jizzakh (9%) and one of Kashkadarja province — covering a total irrigated area of 590,000 ha (12–13% of the national area). The re-designed BISA boundaries are based on hydrographic principles, according to the decree and the international donors' rationale, differing from other BISAs which were finally established on administrative principles (Fergana, Andijan, and others).[353] The three districts of Jizzakh province and the district of the Kashkadarja administrative unit were included according to the Zeravshan valley canals' system, because Tuy-tartar and Eski-Anghor canals respectively carry the water of Zeravshan river to those neighboring provinces. However, as it was stated by local experts and BISA members, the reasons behind the re-designing of the basin agency's boundaries were rather more political than related to the implementation of the reforms according to the IWRM rationale. The boundaries of Zeravshan BISA were finally designed in 2002–2003 by the staff of the former Samarkand province water department.[354] Furthermore, the 2001 proposed BISA (that is, the entire Zeravshan valley including Amu-Bukhara pumping station) was totally distorted by the province authorities' idea which was in conflict with the Ministry of Agriculture and Water Resources. In fact, this re-centralization of basin water management under the Minister's control was not appreciated by the province governments; it would have been challenging for the three former water departments to merge, or to create a new management authority, particularly regarding their bureaucracies and related powers. Therefore, two divided BISAs were created and the Zeravshan BISA, according to the data collected, seems to have decided to include the neighboring districts in order to be based on hydrographic principles in compliance with the decree.[355] Furthermore, Zeravshan BISA members stated that the staff has not significantly changed compared to the former province water department's staff. Through the decree, the Ministry of Agriculture and Water Resources formally accepted these territorial and organizational changes, although they were decided by the local governments.

---

[353] IWMI, 2012."cit.".

[354] Personal communication with Zeravshan BISA' members, Samarkand, May 2011.

[355] Personal communication with Zeravshan BISA' members and local experts, Samarkand, April 2011 and 2012.

### 5.2.2 The uthopic move towards a participatory approach and bottom-up practices

Besides the territorial characteristics (hydrographic principles), another major pillar supported by the donors' rationale concerning the BISAs establishment was the creation of a governance structure, characterized by a participatory approach, according to the model of the basin councils. In the ten years since the Zeravshan Basin Agency's creation, no governance structures have been yet established; in Decree n.320 of 2003, the governance principles were not formally mentioned, hence the idea of setting up the councils should have been promoted by the BISAs' heads or directly by the Minister. Even though no official support from the Ministry has emerged for ten years, the other Zeravshan BISAs' governing boards also have not promoted any basin councils or meeting sessions. In addition, as will be further discussed, both the members and the water users have not induced the governing board to create these types of structures featured by a participatory approach. According to a BISA worker, "organizing meetings and councils among the water users is too complicated; the water allocation process is carried outwell by our authority, hence, there is no sense in changing the organizational processes". Furthermore, another member — a hydro technician working in the former water department since the Soviet period — claimed that the councils were supported, and in some agencies established, in the regions where international projects were designed by the donors — for instance, in the Fergana Valley — but that in the Zeravshan BISA, members and users are able to communicate and solve organizational and technical issues through informal meetings.[356] Therefore, from the data it emerged that regarding the organizational structure, Zeravshan BISA retained a vertical/top-down approach inherited from the former province water department. In addition, no significant changes concerning the staff and, therefore their rationale, have emerged in the last years. Besides the approach in the decision-making processes, the tasksalso have not significantly changed: the basin agency is responsible of the operation and maintenance of the main water infrastructures of the basin, as primary level canals and reservoirs. The canals under BISA's Main Canals System Authority are nine: Dargom, Obvodnoi Dargom, Anghor, Eski-Anghor, KPC, and Narpai in the southern branch, and Bulungur, Pai, and Tuy-tartar in the northern one. Furthermore, the Zeravshan BISA, as the other agencies receive a total water quota *(limit)* from the Ministry of Agriculture and Water Resources, which has

---

[356] Personal communication with Zeravshan BISA' workers, Samarkand, April 2012.

to be divided among the sub-departments, the Irrigation System Authorities (ISAs) formalized within the 2003 decree. As analysed in depth in the previous chapter, the established ISAs throughout Uzbekistan were mostly created according to the territories of the former district water departments (*Rayvodkhoz)*,although in the decree no formal measures were enacted regarding their boundaries. Therefore, most of the ISAs were restructured, according to organizational and operational tasks, similarly to the former departments. Reflecting on the data, a deviance from the mainstream Uzbek reform plan emerged: within this basin agency, the ISAs were designed and organized *ex-novo* by the BISAs' governing board in 2003, as stressed by the interviewed employees. Although they would have to reorganize the former district water departments into the ISAs, they independently decided to set up the new sub-departments according to a local initiative.[357] The Ministry of Agriculture and Water Resources did not express any comments when the Zeravshan BISA presented the plan to the national authorities. Therefore, eight ISAs were newly established: Dargom, Eski, Tuyatartar-Kli, Mirzapai, Miankal-Toss, Narpai, Karmona-Kanimex, and Ak-Karadarja. Questioning the BISA's members about which parameters the Irrigation System Authorities had designed, they argued that most of them were set up according to the main canals and the secondary canal, and their irrigated areas (FIG. 15).[358] Analysing the ISAs' territorial features, it emerged that these parameters were interpreted in different ways; some of them are based on hydrographic principles, others on physical-territorial features or designed covering parts of different districts. Dargom ISA includes three districts (Urgut, Tailok, Pastdargom, and part of Samarkand and Nurabad) according to the canal Dargom; Eski ISA includes part of Nurabad district and Chirokci district in Kashkadarja province, following the course of Eski-Anghor canal, while Karmona-Kaminex and Miankal-toss ISAs — which until 2003 was part of the Navoi province water department's territory — are today included in Zeravshan BISA and refer to secondary canals and small watersheds.Although the ISAs can be considered new organizations, in particular regarding physical and territorial concerns, most of the current staff includes the employees of the former province and district water departments. In certain cases, one employee, as stated in Dargom ISA, can work both in the BISA and ISA — for instance, a hydro technician — or in ISA and in one WUA.

---

[357] Personal communication with Zeravshan BISA' members, Samarkand, April 2012.
[358] Personal communication with Zeravshan BISA' members, Samarkand, October 2012.

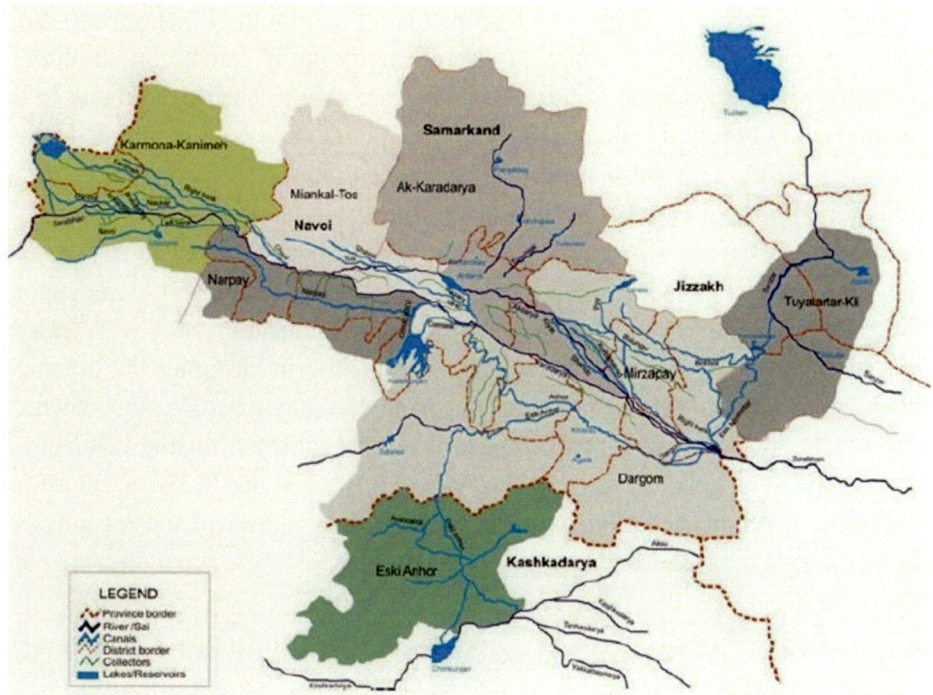

*FIG. 15: Thematic map representing the Zeravshan BISA' territory, the subdivision of ISAs' one and the irrigation network; (source: Wegerich, 2014, approx. scale 1:2.000.000, N↑).*

The interviewed members claimed that their tasks have not significantly changed since the water management's reorganization in 2003. When asking them about the reasons behind the Zeravshan BISA's organizational features — meaning that the ISAs do not refer to the former district water department's territories — they argued that it was a local initiative sponsored by BISA, related with WUAs' features, without going into depth about the rationale; however they stated that the organizational framework of Zeravshan BISA works fairly.[1] The basin agency — once received the water quota from the MAWR — divides it among all the ISAs which are liable for the operation and maintenance of the secondary canal network and for the water allocation to the water users associations, depending on their specific water request.[2] The interviewed *miraab* added that due to the upstream position in the Zeravshan valley's canal system

---

[359] Personal communication with Dargom ISA members, Urgut, October 2012.

[360] Personal communication with Dargom ISA members, Urgut district, October 2012.

and a fair management of the water infrastructures, from the 1st May Dam to the secondary canals level, problems of water scarcity rarely emerge; in addition, if technical or organizational issues occur, they are able to handle them due to the relations with both Zeravshan BISA and WUAs' members. As mentioned above regarding BISA, ISAs also have not established any sort of governance structures supporting a participatory approach in the decision-making processes. According to a Dargom ISA employee, "organizing meeting and councils among the water users is too complicated and many farmers probably would not support these kinds of activities"; he added that the water allocation process is carried out fairly by the authorities, hence there is no sense in changing the organizational structure.[361] Therefore the evidence showed that a vertical top-down organizational structure still characterizes water management at the basin level, both in the inner relations among Zeravshan BISA and the ISAs and those between the ISAs and the WUAs, despite the formal support of the reforms oriented towards the IWRM.

## 5.3 TOWARDS THE IWRM AT THE LOCAL LEVEL: EVIDENCE FROM THE DISTRICTS

Having analyzed the water management context in the Zeravshan river basin, highlighting the institutional reforms and related issues, in this paragraph the focus will be on the three districts — Urgut, Nurabad and Pastdargom — selected for field research here to highlight and point out the IWRM implementation processes and related issues at the local level. As it was mentioned in Chapter 2, the administrative units were selected according to their territorial characteristics and their physical position in relation to the Zeravshan river and its canals system. Urgut district lies in the upstream part of the valley and is part of the irrigation scheme close to the Chakylayan foothills, while the Nurabad unit is located in the downstream section of the middle valley, in a peripheral position regarding the canals characterized by the steppes. Pastdargom district lies almost in the central section of the Zeravshan's irrigated area (FIG.16).

---

[361] Personal communication with Dargom ISA members, Urgut, October 2012.

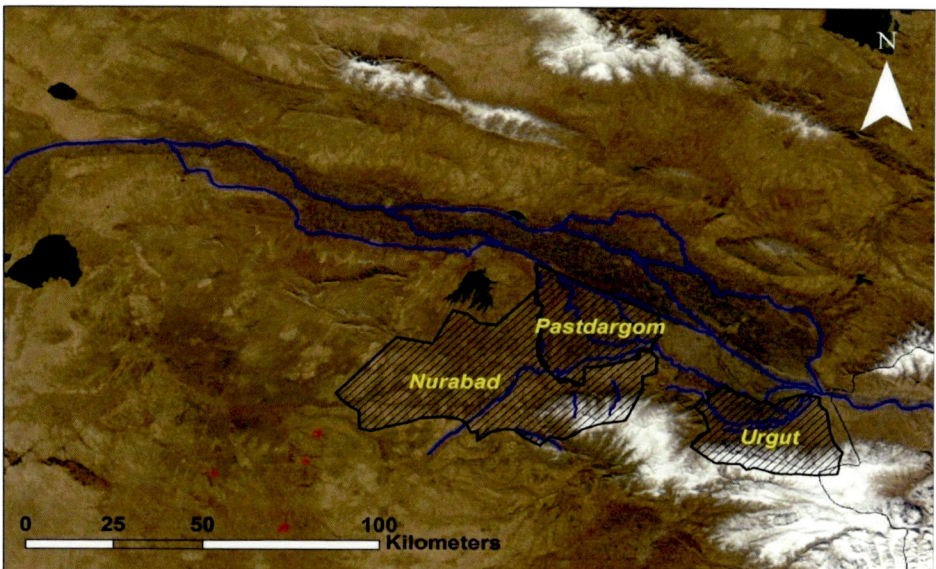

*FIG. 16: GIS elaboration of a satellite image (NASA-Modis, 2003) representing the Zeravshan canals' system (blue lines) and the three administrative units (black shapes) selected for the field-research.*

Furthermore, throughout the districts, some villages were selected for fieldwork in order to highlight and understand the water management issues at the farm level. As previously analysed and discussed, the reorganization of the water sectorat the local level in Uzbekistan — in accordance with the IWRM framework and the IMT — has been in progress since the 2000s and was formalized in the 2002 decree supporting the establishment of the Water Users Associations.Therefore these processes will be outlined and debated, focusing in particular on the institutional, organizational, and operational features of the water authorities, basing the analysis on the IWRM pillars: WUAs' organizational structures, territorial features (administrative and hydrographic), governance procedures and ISF. The analysis will be integrated with data collected through interviews and informal talks with the farmers regarding the WUAs' performance and the potential local water management lacks and issues.

### 5.3.1 URGUT DISTRICT

As shown in the map above, Urgut district lies on the upstream side of the Middle Zeravshan valley (south-eastern part of the Samarkand province) near the boarder with Tajikistan, in the foothills area of the Chakylayan mountains (Zeravshan range, 2616 meters a.s.l. the highest peak) which divides the Zeravshan

148

basin from Kaskadarja's.[362] Urgut, the chief town of the district, is located almost at the centre of the administrative unit on the alluvial fan (900–1000 m. a.s.l.) built by the homonymous river. Urgut district can be physically divided into three parts: the mountain area in the south, the foothills area built by the small rivers arising from the mountain range, and the alluvial plain. The altitude ranges from 2600 a.s.l. in the S to 720 metres a.s.l. in the NW.[363] The central-northern part of the administrative unit is entirely crossed by the main canals of the southern branch of the Zeravshan canals system, arising from the 1st May Dam: Obvodnoi Dargom, Yangiarik, and Mashini canals (FIG. 17 and 18).

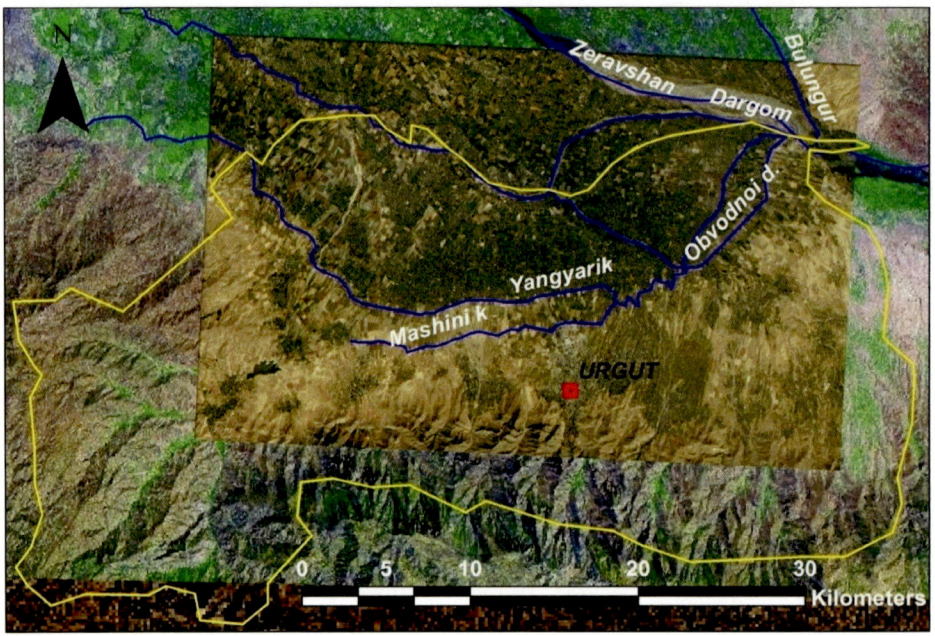

FIG. 17: GIS elaboration of a patchwork of satellite images (Geocover, 2000/Landsat 7) representing Urgut district (yellow shape), its territorial features and the canals system (blu lines).

Since these water facilities are of primary and secondary levels, they are managed and controlled by the main canals systems' authority of Zeravshan BISA (Dargom and Obvodnoi Dargom) and by Dargom ISA (Yangiarik and Mashini).[364]

---

[362] BENSIDOUN, S. 1979. "cit.".

[363] FEDCHENKO, F. 1871. "cit.".

[364] Personal communication with Zeravshan BISA and Dargom ISA members, Samarkand and Urgut,

## *Urgut WUA: the evidence of local political initiatives*

The district's whole irrigated area reaches 30.615 ha, and focusing briefly on the average annual rainfall, it ranges from 350–400 mm/y in the northern part of the district and 400–450 mm/y in Urgut, to 800 mm/y in the Chakilayan mountain range.[365]

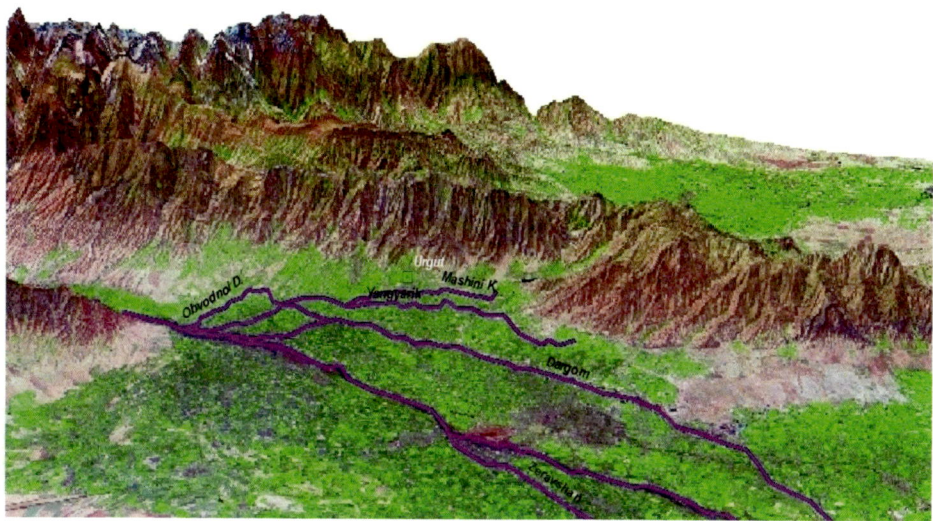

*FIG. 18: GIS elaboration of a 3D satellite image, 2009, (Geocover 2000, source: Uzbek-Italian archeological project, University of Bologna) representing the Middle Zeravshan valley and its canals' systems (blue lines) and Urgut (green point), oriented from N. to S. Approx. scale: 1:1.500.000.*

According to the national reforms for the water sector at the basin/local level issued in the 2000s, the district water department (*Urgutski Rayvodkhoz*), a branch of the Samarkand province water authority operating since the Soviet Union, was officially dismantled at the end of 2002.[366] Although it has not been formalized by decree, throughout Uzbekistan the WUAs have been organized on the basis of the former *shirkat* and their territories, measuring on average 2000–

---

April-October 2012.

[365] BENSIDOUN, S. 1979. "cit.".

[366] Personal communication with Zeravshan BISA members, Samarkand, april 2011.

3000 ha.[367] In the Urgut district, the evidence from the first data has shown a significant deviance from the mainstream Uzbek reforms at the local level. Although the Urgut WUA was established in 2003 by order of the district government (*Hokimyat*) as a non-profit organization and registered in the Justice Department according to governmental decree, the WUA was organized on the basis of the administrative boundaries of the entire district; therefore the water association refers to the territory of the former district water department (*Rayvodkhoz*). Hence, no changes regarding either boundaries or territory have been carried out since the collapse of the Soviet Union.[368] In addition, although the WUAs should be organized by the water users, in a bottom-up perspective, according to the donors' rationale, Urgut WUA was established by the district and local water bureaucracies. Urgut WUA's irrigated area reaches a total of 30.400 ha and is divided between the land owned by the state and leased to the peasant farmers, and private household plots (*tomorka*): the farmers' leased land (545 registered in the district *hokimyat*) accounts for 24.200 ha while the remaining 6.200 are private plots. Most of this land is irrigated by gravity but, regarding the territory's morphology, 7.500 ha are supplied by the 22 pumping stations built since the 1960s and controlled by the province and district branches of the Ministry of Energy. Therefore, on average, the Urgut WUA is approximately five times bigger in size than the average WUA in the other Uzbek provinces; in this district the water association controls the irrigated area, which in the other basins is under the supervision of the ISAs,[369] According to the Urgut WUA's director, the association was established following the organizational principles "one district, one WUA", despite of the measure issued at the national level oriented towards the IWRM.[370] He argued that water management in the Urgut WUA is much better compared to the small WUAs: if some data is requested by ISA, they are able to get it easily, as it is one WUA; while in the other smaller associations, the staff cannot collect data on time or correctly. The staff includes the director, one main hydro technician, one accountant and 24 *miraab*; each of them (*miraab*), despite working for the WUA, is related, in terms of their tasks, to the irrigated area of the *agrofirma,* which measures on average 1500–2500 ha.[371] As the Urgut WUA is entirely located in the territory of Dargom ISA, the

---

[367] Personal communication with GIZ experts, Tashkent, April 2012.

[368] Personal communication with the Urgut WUA's director, Urgut, April 2011-2012.

[369] Personal communication with IWMI experts, Tashkent, October 2012.

[370] Personal communication with the Urgut WUA's director, Urgut, April 2011-2012

[371] The territory of Urgut WUA is nowadays subdivided into 24 *agrofirma;* those associations were

staff, once collected the farmers' crops plan, prepares the documents focusing on the water request for the whole cropping season to be submitted to the Irrigation System Authority. This procedure is facilitated by the fact that the WUA director is simultaneously a member of the staff of Dargom ISA and is specifically dealing with Yangiarik canal's operation and maintenance.[372] The Urgut WUA is therefore responsible for water allocation to the farmers at the tertiary canals level; these tasks are up to the *miraab*, which provide water in turn according to the WUAs' schedules. However at the farm level, there are not any hydro posts or metal gates, hence, it is not possible for the *miraab* to exactly calculate the water volume, so they just use mud and soil to open and close the outlets and to divide the water flow.[373]

### The ISF conflict with the WUA's logics

Focusing on the water users, the Urgut WUA only includes the large farmers (545 in 2012) engaged with wheat, tobacco and grape farming, since, due to the soil's features in this district, cotton cropping is not widespread. the farmers' lands nowadays range on average from 20 to 70 hectares since the optimization process *(optimisazija)*, started in 2009, which has reduced the number of farmers, thus, increasing the land under their control.[374] At the same time, the household plots' owners are not involved in the WUA, even though the concept of integration of all water users and different water uses is one of the main pillars of the IWRM, strongly supported by donors. These farmers stressed that they are not integrated in the WUA because it involves just profitable agricultural landowners (wheat and cotton farmers), according to the *Hokimyat* demands; nevertheless they receive water from WUA's *miraab* directly to their plots for free, which also conflicts with the reforms' rationale and the ISF.[375] In fact, as highlighted in the previous chapter, the Irrigation Service Fee (ISF) has not been officially formalized by any measures at the national level; it has been only men-

---

the *kolkhoz* during the Soviet Union, afterward they shifted into the *shirkat* until 2006-2007 when the dismantling process terminated leading the rise of peasant farmers. Nowadays *agrofirma* are State Enterprises responsible of farm's productivity and to provide services for the farmers as fertilizers, machines and tractors; furthermore, according to the State quota for the main crops, cotton and wheat, they have to ensure the *hokimiyat'* crops demands.

[372] Personal communication with the Urgut WUA' sstaff, Urgut, October 2012.

[373] Informal talks with Urgut WUA' miraab, Yangiarik canal, October 2012.

[374] Personal communication with GIZ' experts, Tashkent, april 2012 and Urgut WUA' staff, Urgut, April 2011.

[375] Personal communication with the household plots, Urgut district, April and October 2012.

tioned in the decree on WUAs and sponsored by the international donors as an indispensable tool for the WUAs' financial availability. According to the Urgut WUA's director, the ISF was introduced in their WUA in 2004, but in the first years the *shirkat* still functioned and so they did not pay for water allocation services; the same was true for the first peasant farmers who were not used to paying water charges. The director argued that since 2009 the water supply charges have been widely applied to the farmers, although some of them do not pay or are not able to pay; the annual fee for water service, measured in hectares, since it is impossible to account for the water flow in cube meters, is approximately 16,000 sum/ha (6/7 USD) [376]. The Urgut WUA's director did not mention if the collected fees are adequate to cover the operation and maintenance expenses. This lack of information leads us to assume that the WUA receives financial support from the district state authority, otherwise, it would be impossible for the association to operate. The farmers interviewed in the villages argued that some of them do not pay water service charges because of money shortages due to other expenditures, such as seeds, fertilizers, and tractor rental. They claimed that since the farmers have made an agreement with the government for cotton and wheat farming, even if they are not able to pay the WUA, they would receive the water amount requested nevertheless, since they are supported by the *Hokimyat*. [377] This evidence is in sharp contrast with the IWRM rationale, making a clear understanding of the governmental logics affecting the WUAs.

### *Participation does not rise: maintaining a top-down approach*

Focusing on the participatory approach in the decision-making processes strongly promoted as a necessary tool for the IMT, it is apparent that a real governance structure within the Urgut WUA has not been yet organized. According to the WUAs' staff, informal councils in the last years have been arranged (consisting of 3 to 5 members) but the meetings are rarely scheduled and weakly supported; in addition, Zeravshan BISA has not promoted the establishment of a governance board and the WUA's staff has not expressed any intention to increase the participation of the water users in the decision-making processes.[378] It would be extremely challenging to establish a governance structure, including

---

[376]Personal communication with the Urgut WUA' director, Urgut, October 2012.

[377]Personal communication with the farmers, Urgut district, april and october 2012.

[378] Personal communication with the Urgut WUA's staff, Urgut, October 2012.

councils, considering the very large number of water users (545 farmers). According to the data collected, many of them have never expressed any desire to participate in the WUA's decision-making processes for multiple reasons — saying, for instance, it is not their job to do so as farmers, or that they are afraid to involve themselves in organizational, and somewhat political, issues; other farmers claimed that they have not been informed by the WUA's staff about meeting and councils.[379] Furthermore, as emerged from the interviews with both experts and WUA's members, since the Urgut WUA establishment in 2003, no elections were organized to decide the head of the association who was directly appointed by the district *Hokimyat,* limiting any kind of participation of the water users in the decision-making procedures. As mentioned in the previous chapter, in 2009, Law n.240 leading to amendments to the 1993 Water Code, was enacted by the government with the aim of strengthening the WUA's role in water and environmental sustainability; WUAs based on hydrographic principles as well as the active participation of the water users were mentioned.[380] Notwithstanding this measure, in the last three years, has not lead to changes in Urgut WUA's institutional and organizational framework. As claimed by the members, some of them only heard about this measure issued at the national level and did not receive any official documents concerning the changes; in addition, any request from Zeravshan BISA was made to change the WUA status and to shift to hydrographic boundaries.[381] Furthermore, to be based on hydrographic principles, the WUA should be split into several smaller associations. About the national measure, the WUA's director claimed that they are able to achieve their results, providing fair water management and allocation to the water users, throughout their territory; therefore he argued that they do not have any reason to change the WUA's organizational framework. He also added that they work in close relation with Dargom ISA, and that it would be very challenging to split the associations in part due to a potential lack of hydro technicians. Asked about whether in the near future they will be obliged by the government to implement the law and establish WUAs according to hydrographic principles, the WUAs' director argued that they were able to design the associations in 2003 based on district boundaries, hence no changes will occur.[382] As mentioned above, the territorial and organizational features of Urgut WUA represent a deviance from

---

[379] Personal communication with the farmers, Urgut district, april and october 2012
[380] IWMI, 2012."cit.".
[381] Personal communication with the Urgut WUA's staff, Urgut, October 2012.
[382] Personal communication with the Urgut WUA's director, Urgut, October 2012.

the mainstream reforms. According to the director, the organizational principle of "one district, one WUA" was a local initiative, widespread in most of the Samarkand province's districts, decided and carried out by the old Samarkand province water department's members (*Oblastvodkhoz*). This was a decision undertaken by "high level" authorities, in close connection with the Ministry of Agriculture and Water Resources, therefore, nobody wondered whether it was a deviation from mainstream reforms, or whether, with these characteristics, this local initiative does not implement the IWRM pillars and is far from its rationale.

### The everyday water practices in Urgut WUA's villages

In order to go in- depth in the analysis of farm-level water management issues, the evidence from three villages are presented below (FIG. 19):

*Tegana village is located almost at the centre of the district in the irrigated alluvial plain at an altitude of 850 metres a.s.l.. The main irrigated area oriented towards agriculture lies between Yangiarik canal in the north and Mashini canal in the south. The Tegana farmers' irrigated area reaches 700– 800 ha; the village was included in kolchoz Ilich until 1991 (1900 ha, four villages), then in shirkat Yangiarik until 2006, when it was dismantled. The lands were divided in 2007 among 84 farmers, and, nowadays, after the Optimisation process started in 2009, among 24; on average each farmer has 30–40 ha of leased land. The main crops cultivated on Tegana lands are wheat and tobacco, fruit, and orchards. Farmlands are irrigated by gravity and flood irrigation from Mashini canal which is powered from May to October by five pumping systems, while in the other months it receives water from mountain rivers in particular from Urgutsai. Water allocation at the tertiary level in Tegana village is up to WUA miraab, one or two depending on the months: according to the household plots owners, water allocation is fairly and equally provided, preventing disputes among them. They claimed that water shortages in the village do not occur because of the permanent flow of Yangiarik canal. Household plot owners do not pay charges for water allocation, they just pay an annual fee in land taxes. Participation and integration pillars are not widespread at the farm level. Household plots owners living in Tegana are not integrated and included in Urgut WUA. Most of them still know the WUA as Rayvodkhoz and state that the association includes only the large farmers who are involved with wheat and tobacco farming. Due to the favourable position of the village and the proximity to the main canals of the irrigation network, the large farmers do not face water shortages during the cropping season: in springtime Chakilayan snow melting and in summer Zeravshan ice melting provide a sufficient amount of water to achieve the plan stipulated with the government. Therefore, despite the large number of both large farmers and household plot owners, the Urgut WUA ensures fair water allocation to all the water users, provides the operation and maintenance of the canals at farm level, avoiding potential disputes and issues among the farmers. Therefore, all the water users claimed that despite the large surface of the WUA and the large number of farmers involved, the organization properly works, hence, they do not expect any changes in the institutional and organizational framework.*

*Sariktepa village lies in the south-western branch of Urgut district in the Chakilayan western foothills area. Regarding the canals system, the village lies between the Yangyarik canal course in the north and Mashini canal in the south. The water management dynamics in Sariktepa village greatly differ, from a technical perspective, from Tegana's. The village water supply is ensured by both gravity and pumping irrigation: one pumping station lifts water from Yangyarik canal two to three kilometres north of the village where it subsequently flows down through a tertiary level canals. In addition the Mashini canal provides water only from May to September. Saryktepa is often affected by water shortage issues because, due to the foothills' position higher than the Yangiarik canal's course , gravity irrigation is not possible. In addition, since the village lies at the tail end of Mashini canal, unavailability of water is frequent. The lack of energy is an everyday challenge in Sariktepa, hence, the pumping system lifting water from Yangiarik rarely works; the water infrastructure is controlled by the district department of energy, but, according to the household plots owners, due to lack of finances, it is not fairly maintained. According to the household plot owners, in some months, especially in summer, they collect money through the village authority (mahalla) to pay for the energy to run the pumping station. During the cropping season, most of the Mashini canal's water availability is used for the large farmers' land irrigation (wheat and tobacco), and consequently, the kitchen gardeners are excluded from this water source. Moreover, in Saryktepa there are not any relations between the household plot owners and Urgut WUA: water allocation at the farm level is carried out by one to two miraab of the village who have not ever worked in either the WUA or in the former Rayvodkhoz. They are not hydro technicians; they just divide and deliver water among the users as a service to the community — sometimes for free and sometimes receiving a small fee. Therefore in Sariktepa a relevant discrepancy, differently from Tegana, emerged among the large farmers, whom water services are ensured by the Urgut WUA, and the small plot' owners, which had to face with water shortages' issues organizing themselves an informal service.*

***Jarkishlak village** lies in the eastern side of Urgut district near the border with Tajikistan in the eastern part of the Chakilayan foothills (altitude 920 metres a.s.l.). The village has approximately 5000 inhabitants and was included, until 2006, in Akkurgan shirkat. The main crops cultivated by the farmers are tobacco, grapes, and wheat. Since the village lies higher than the Zeravshan canals network, water supply in Jarkishlak is ensured by the state pumping station, Dargom 1, lifting water 100 metres higher and 5000 metres in length, irrigating 2000 ha. At the top of the pumping system (altitude 1005 metres a.s.l.), in the fields between Jarkishlak and Muminabad, the water is delivered to the village through three secondary canals and a network of tertiary canals. Those canals are controlled and maintained by the Urgut WUAs' mirabs which, as argued both by the farmers and the household plot owners, work every day, overseeing the outlets and dividing the water flows according to the water users' schedules. The interviewed farmers claimed that generally they are not affected by water shortages in the village; when some problems emerge, rarely, they are strictly connected withlacks of energy. This issue is not frequent, in contrast with Sarikyepa, since Urgut 1 is the most important pumping station of the district. Nevertheless, in the cases of water unavailability, the large farmers are the first to receive water allocation service, as they are involved with state crops. They claimed that no relevant changes occurred since the shift from Rayvodkhoz and Urgut WUA except that they have to pay the water charges according to ISF, but only few of them actually pay. Regarding the small plot owners, water is delivered for free (they just pay annual land taxes) and they are not integrated or included in Urgut WUA, as emerged in Tegana. Nevertheless, they do not suggest any kind of changes in farm level water management because, as they argued, Urgut WUA works properly, avoiding potential disputes among them.*

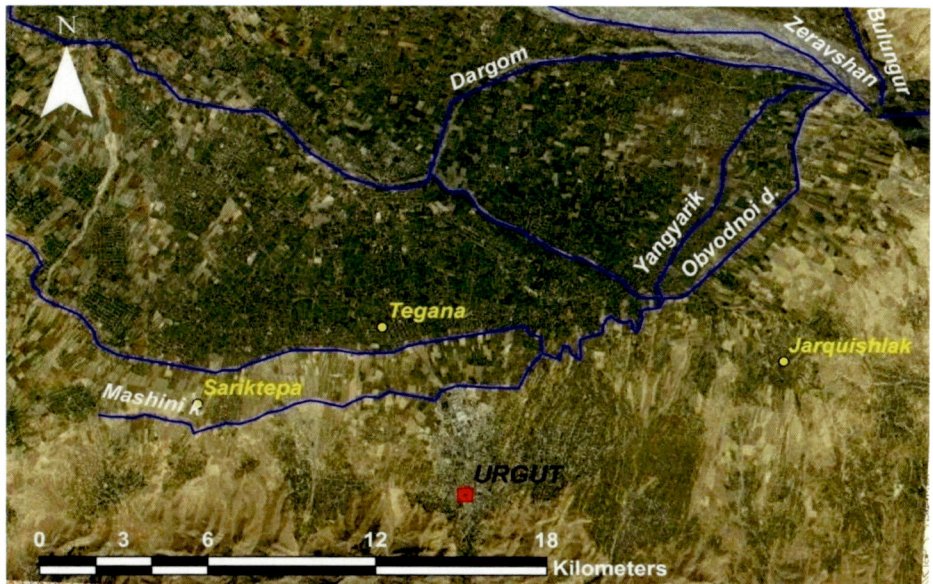

*FIG. 19: GIS elaboration of a satellite image (Google Earth TM) representing Urgut WUA's territory and the three villages investigated (black points, names in yellow) (approx. scale 1:300.000, N↑).*

### 5.3.2 NURABAD DISTRICT

Nurabad district lies in the western-southern side of the Middle Zeravshan valley in a peripheral area relative to the valley's irrigation scheme arising from the 1st May Dam; due to its geographic position its large territory (almost three times Urgut district' area) is mostly characterized by the steppe area named Nurabadciul – Karnabciul which divides the Zeravshan catchment from the Kashkadarja that lies to the south.[383] According to the territorial features, Nurabad district can be divided into three areas: the eastern one, featured by the Karatyube range (highest peak, 2224 m. a.s.l.) and its foothills; the central one, crossed by the two main water courses (Sabir river and Eski-anghor canal), where the irrigated areas are located; and the western one featured by the steppe and by low dry hills.[384] Due to the physical-environmental conditions, Nurabad district's irrigated area reaches a total of 6.088 ha, less than 10% of the whole territory and it is supplied by the Eski-Anghor canal and Sabir river (FIG.20).

---

[383] BENSIDOUN, S., 1979. "cit.".

[384] ZINZANI, A., 2011. "cit.".

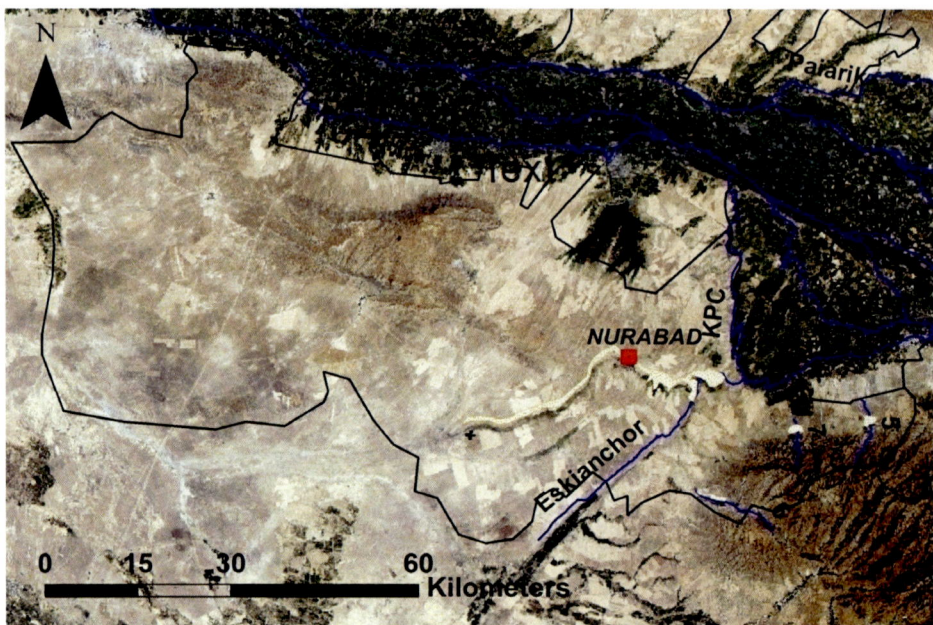

*FIG. 20: GIS elaboration of a satellite image (Landsat 5) representing Nurabad district (black shape), Nurabad town (red point) and its irrigated areas (white shapes) (approx. scale 1:1.000.000, N↑).*

This river flows for 40 km, supplying Nurabad town and terminates into the ho-monymous small reservoir; the district's largest irrigated area lies on its banks, where several small canals arise, and its width  approximately ranges between one and five kilometres.[385] Other small irrigated areas lie in the northern part of the district near the village of Nurbulak, supplied by gravity from the KPC ca-nal, and in the north-western area on the border with Samarkand province irri-gated by the Narpai canal. Considering the physical features, the low annual rainfall and the distance from the main water sources, in Nurabad district, since ancient times, livestock breeding and household plot farming prevailed in com-parison to extensive agriculture. Irrigated agriculture began to be practiced only a few decades ago: until the 1980s  this area was used as grazing land for the *sovchoz* involved with livestock and fodder farming.[386] Irrigated wheat and cot-ton farming were introduced during the end of the 1980s and in the 1990s, de-

---

[385] Personal communication with Ulus' major, Nurabad province, April 2012.
[386] BENSIDOUN, S., 1979."cit.".

spite the quality of the soil (brown loam soil) and frequent water scarcity issues.[387]

### Nurabad WUA: maintaining the local political initiatives

As analysed in the Urgut unit, in Nurabad, the district water department (*Nurabadski Rayvodkhoz*) was dismantled at the end of 2002 according to the national law. The evidence has shown other notable similarities with the Urgut district, focusing on the WUA establishment; also in Nurabad, the deviance from the mainstream national reform, decided and carried out by the Samarkand province water department' members before its reorganization into Zeravshan BISA, occurred. Therefore, any territorial changes with respect to the *Rayvodkhoz* area was supported; the Nurabad WUA was established in 2003, based on administrative principles, according to the district boundaries.[388] Therefore, as in Urgut district, a local interpretation of the institutional reform emerged. Regarding its institutional framework, Nurabad WUA was created by the order of the district government *(Hokimyat)*, hence not by the water users, in contrast with the IMT rationale, and registered in the justice department. Since the territory under the WUAs' control measures only 6088 ha, it is not so different in sizein comparison to the associations organized throughout Uzbekistan, measuring on average 2000–2500 ha; nevertheless, in Nurabad district, the irrigated areas lies far from each other and are supplied by three different canals (Eski-Anghor/Sabir, KPC, and Narpai), making the water control and allocation more complex and challenging. According to the Nurabad WUA's director, due to the small irrigated area involved, they could retain the same territory and boundaries of the former district water department; this statement assumes an agreement with the Samarkand province water department, although the local initiative regarding the WUAs' establishment in Samarkand province was not mentioned. As emerged in Urgut WUA, in Nurabad the director was appointed directly by the *hokimyat* in 2003, and in the last decade no changes at the head of the governing board have occurred. The board also includes one hydro technician, an accountant, and six *miraab* which, since the Soviet Union, have been members of the Nurabad district water department. Regarding the relations with the Zeravshan ISA, concerning the water allocation demand, the evidence has shown a significant difference with Urgut WUAs' organizational and operational structure; due to the

---

[387] Personal communication with District Agricultural department, Nurabad province, April 2011.
[388] Personal communication with Nurabad WUA' director, Nurabad, April 2011.

large territory and three irrigated areas lying far from each other, Nurabad WUA refers annual water requests to three different ISAs: Eski ISA for the eastern side, Dargom ISA for the north-eastern side and Narpai ISA for the north-western one. According to the WUA director, the reorganization in 2003 of the Samarkand province's water department (the switch to BISA) and the subsequent establishment of the ISAs, based on new boundaries, strongly affected organizational issues; the head of the governing board added that this condition led to lacks concerning water supply procedures due to the more complex organizational network and the demand for separated water between the WUA and the three different ISAs.[389] Concerning the water users, Nurabad WUA nowadays includes a total of 65 farmers which are involved with wheat and cotton farming .[390] In the Urgut WUA, even though the household plots owners are not included and integrated in the association, they receive water from the WUAs' *miraab,* *whereas* in Nurabad the water allocation in the villages is carried out directly by the plotowners themselves, without any contact with the WUA. As argued by the Ulus village mayor, the idea of integrating the small plot owners in the WUA has never been discussed; he stresses that as the WUA is still controlled by the local water bureaucracies, it is interested only in supplying water to the farmers who cultivate state crops.

### Few changes in governance: the WUA organizational structure

Despite the small number of these farmers included in the association, which would make possible to organize meetings and appoint farmers' representatives, any official council has been sponsored by the WUA's governing board. Although the director admitted that sometimes in the district water allocation issues occur, mostly due to water scarcity, they are able to face with those problems themselves without the involvement of the water users[391]. Furthermore, the farmers interviewed stated that it is sufficient to meet the WUA's staff twice a year, at the beginning and at the end of the cropping season, and they added that more meetings are not useful for them; they claimed that whether some water

---

[389] Personal communication with Nurabad WUA' director, Nurabad, April 2011.

[390] Their land leased from the state ranges on average from 60 to 90 ha, larger in size in comparison with the Urgut WUA due to the territorial features and to the lower population' density. According to the head of the District Agricultural Department (*Rayselkhoz*), before 2009 when the land' optimization process (*optimisazija*) started, the farmers were 90, cultivating approximately 50-60 ha.

[391] Personal communication with the Nurabad WUA' director, Nurabad, April 2012.

allocation' issue emerge, they directly discuss with the *miraab* on the fields or they go to Nurabad to the WUA office[392]. Different farmers argued that in the last years water unavailability and allocation' mismanagement have been frequent, partly due to the physical position of the district, in case of water shortage they are at the tail-end of the canals' system, and to organizational lacks of the WUA. Although, as occurred in Urgut WUA, the ISF has been widespread only since 2009 (officially introduced since 2004), in the last four years, despite the partial water charges collected from the farmers, the water delivery services have not improved. According to the WUA' accountant, the water fee is approximately 10.000 sum for one ha/year (4.5 USD), cheaper than in Urgut WUA, but several water users do not pay, leading a financial unavailability for operation & maintenance.[393] The farmers interviewed claimed that part of them do not want to pay the water charges since the water delivery' services are not provided according to the schedules; in addition they added that whether after a couple of years of payment, the service does not improve, there is no reason to continue paying the fees.[394] The WUA's director argued that water management was much better when the district water department operated, during the Soviet union, because it was totally supported by the government; moreover the former authority had its own techniques and more funds for operation & maintenance, which made water allocation more suitable. What emerged is a complex context, from one side the WUA do not provide suitable service and is affected by financial shortages and from the other the water users are reluctant to pay for inadequate water allocation' service. Therefore the IMT, as promoted by the reforms, has not been from one side implemented by the local bureaucracies and from the other not supported by the water users. Despite those issues, the 2009' law on WUAs, which through its implementation could lead improvements to the Nurabad WUA's organizational and institutional structure, have not led any changes in the association' status. According to the WUA's staff, they do not receive any communication from the BISA to implement the national measure. Nevertheless at the end of 2011, considering the WUA' organizational issues and in order to face with the analysed lacks, Zeravshan BISA, aware of the Nurabad water management' context, lodged a special request to the director to split the WUA in three new associations based on hydrographic principles. The new

---

[392] Personal communication with the farmers, Nurabad district, April 2011.
[393] Personal communication with Nurabad WUA's accountant, Nurabad, April 2012.
[394] Personal communication with the farmers, Nurabad WUA, April 2012.

WUAs should refer to the three canals and connected irrigated areas (Eski-Anghor, KPC and Narpai) and should be territorially related to the different ISAs (Eski – Dargom-Narpai) [395]. A draft concerning the potential organizational and territorial structure of the new entities was prepared by the staff and submitted to the BISA but, nevertheless, since the last two years, Nurabad WUA has not been yet affected by any institutional and organizational change.

### The everyday water practices in a Nurabad WUA's village

In order to get a clearer overview of the discussed issues at farm-level, one village in Nurabad WUA, Ulus, was investigated:

*Ulus village* *lies in the largest irrigated area of Nurabad district, which measures approximately 2500–3000 ha, supplied by the Sabir river, fed partly from natural wells and partly from the Eski-Anghor canal's flows. Besides these two main courses, this area is irrigated by two secondary canals, ER1 and ER2, designed at the end of the 1970s, originating from the Eski-Anghor. Due to the particular territorial features of Nurabad WUA, those canals are under the supervision of Eski ISA, based in a village in neighboring Kashkadarja province. The Ulus's irrigated area where lie the farmers' plot measures on average 800–1000 ha; concerning the crops, they are approximately equally divided between wheat and cotton farming. After the agricultural optimisation process, the lands have been shared among 12 farmers. According to the farmers interviewed, in Ulus area, water scarcity is a widespread issue both in March and April, when the snowmelting in Chakilayan/Zeravshan mountains begins and hence the Zeravshan flow is scarce, and also in July and August when the Eski-Anghor canal's water flow is reduced because of upstream use. Regarding this issue, one household plot owners argued that in the upstream districts of Urgut, Samarkand, and Pastdargom, the farmers probably use more water than they should. Furthermore, they added that in cases of water scarcity, the irrigation of cotton fields takes priority over wheat fields, due to the importance of this crop for the government. According to the Nurabad WUA issues previously analysed, in Ulus village several water management problems have occurred, in particular in the last two years, both from technical and organizational perspectives. The WUA and its miraab do not properly work and often cannot adequately provide water allocation services to the farmers;*

---

[395] Personal communication with Nurabad WUA' s staff, Nurabad province, October 2012.

*absence of the miraab, lack of maintenance of the canals, and lack of control
of the outlets are everyday issues in Ulus's irrigated area. Therefore, unequal
water distribution, disputes, and the robbery of water between the largest
farmers and the smaller ones is a widespread issue, without any control by the
WUA. Due to this mismanagement, several farmers decided to stop paying water
charges and to organize their own water delivery. Therefore, according to the
farmers interviewed, nowadays the maintenance of the canals is not ensured and
water distribution times are self-organized by the farmers, leading disputes. In
addition, they argued that often the household plot owners divert the water flow
to the village before it reaches the fields. The context in Ulus village is quite
different because the WUA miraab have never worked in plots water delivery,
hence the villagers are able to self-organize preventing disputes; they all meet
together once in March, before the start of the cropping season, and all the
water users participate in the maintenance of tertiary canals inside the village.
As already verified in the other villages — for instance, in Tegana, Sariktepa,
and Jarkishlak in Urgut district — they do not pay water fees. The plot owners
stated that it would be easier if the Nurabad WUA miraabs would help them
with the canals' maintenance, but they know that due to the current
organizational issues, this is not possible. Finally, the evidence has shown that
these management and technical issues are worsening the agricultural
conditions in an area which is already stressed by water scarcity.*

## 5.3.3 PASTDARGOM DISTRICT

Pastdargom district lies entirely within the alluvial plain, almost in the centre of
Middle Zeravshan valley and its canals system, 20 kilometres west of Samarkand; only the southeast part physically belongs to the Karatyube foothills area,
which are characterized by the steppe. This territory is entirely crossed by the
most important canals of Zeravshan's irrigation scheme — Dargom, Anghor,
Eski-Anghor, and KPC which flow in W-E direction — and it is delimited in the
northern side by the Karadarja branch of the Zeravshan river. Since those canals
belong to the primary level of the canals system, concerning operation and
maintenance, they are managed by the Main Canals' Authority of the Zeravshan
BISA.

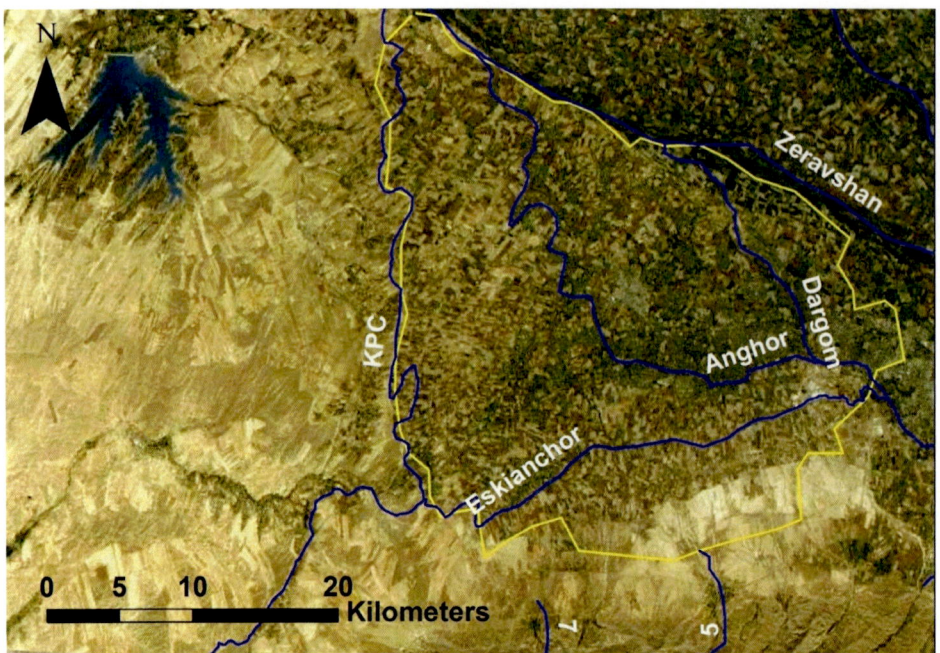

*FIG. 21: GIS elaboration of a satellite image (source: Google Earth TM) repre-senting Pastdargom district, its boundaries (yellow line) and the canals' network (blue lines).*

The Pastdargom district's total irrigated area reaches 53.848 ha, representing one of the more extensive irrigated areas of the Samarkand province; it was divided, after the optimization process, among 1100 farmers, who rent approximately 50–60 hectares from the government (FIG. 21).

### Pastdargom WUA and the donors-based RESP II project: towards a different rationale

The implementation of the 2002 measure on WUAs supporting the IMT has not shown significant differences in comparison with with Urgut and Nurabad dis-trict. According to the local initiative and the deviation from the mainstream re-forms supported by the Samarkand province water department, Pastdargom WUA was created in 2003 by the order of the district *Hokimyat* and based on district administrative boundaries; therefore no changes in territorial features, organizational structure and staff have occurred with respect to the Pastdargom district water department, despite the challenges regarding water allocation to more than 1,000 water users in comparison with the 34 former cooperatives. Nevertheless, according to the WUA director, they were able to manage the land tenure changes quite well, avoiding disputes for water delivery among the peasant farmers and *shirkat* in the first years, and, since 2006, among the farmers.[396] As mentioned in the previous chapter, Pastdargom district, due to its large territory, high population, and suitable agricultural conditions, was chosen in 2008 as a pilot-project area (seven districts within seven provinces in Uzbekistan) for the international project "Rural Enterprise Support Project Phase II" (RESP II) designed and supported by the World Bank and the Swiss Development Cooperation, and implemented by the SIC-ICWC. The project, as already highlighted, aims to improve water use and management at the local level, oriented towards social and environmental sustainability — through the support of the existing WUAs and the establishment of new associations— and towards strengthening the farmers' business.[397] Furthermore, although it is not mentioned in the official papers, the RESP II project addresses the strengthening of the IMT and IWRM rationales, specifically integration and participation' pillars, through the organization of meetings and seminars.[398] According to the evidence, in Pastdargom district, the RESP II project has mainly focused on the institutional and territorial division of the WUA into several associations in order to improve its performance, and on the restructuring of the water facilities at the tertiary canals level. Therefore, since 2009 the particular Zeravshan valley' WUAs context, designed at the beginning of the 2000s by the Samarkand province water department's members, has been affected by structural changes led, for the first time, by the international donors' influence.

### Supporting the hydrographic units: the subdivion of Pastdargom WUA

The main idea of the international organizations has been to design new WUAs based on hydrographic principles, as also sponsored by the 2009 governmental amendments to the water code.According to the staff of the former Pastdargom WUA, the first stage of the project,which started at the beginning of 2009, was oriented towards the division of the district WUA into two different associations: the main one, which is still named Pastdargom WUA and based on the former territory measuring 49.000 ha, and the new one, called Talligulom Meva Uzum WUA, measuring 5094 ha (FIG.23).[399] Despite the decreed and espoused principle of designing new organizations based on hydrographic principles, Talligulom

---

[396] Personal communication with Pastdargom WUA' sdirector, Juma, April 2012.
[397] WORLD BANK, 2008. "cit.".
[398] Personal communication with SDC' sexperts, Tashkent, April 2012.
[399] Personal communication with the former Pastdargom WUA's staff, Juma, April 2012.

Meva Uzum WUA was established according to the administrative boundaries and territory of the former *sovkhoz* Kasimov, as most of the WUAs created throughout Uzbekistan since 2002. Those institutional and organizational changes as well as the design of the boundaries of the new WUAs were decided through meetings among the stakeholders, SIC, former Pastdargom WUA staff and Zeravshan BISA, and finally approved by the donors.[400] The Talligulom Meva Uzum WUA — totally independent after the shift from the former Past-dargom WUA — nowadays regarding water amount request is directly related to Dargom ISA as the other WUAs. The new staff includes the director and the former members, hydro technicians, accountant and four *miraab*, of the *sovkhoz* Kasimov. The director claimed that although since 2003 the former Pastdargom WUA has provided fair water delivery to the water users, the current new institutional and organizational status enables more strictly relations among the water users, avoiding potential disputes among them, due to a smaller number of farmers and a smaller territory (5094 hectares, instead of 54000).

---

[400] Personal communication with former Pastdargom WUA's staff, Juma, April 2012.

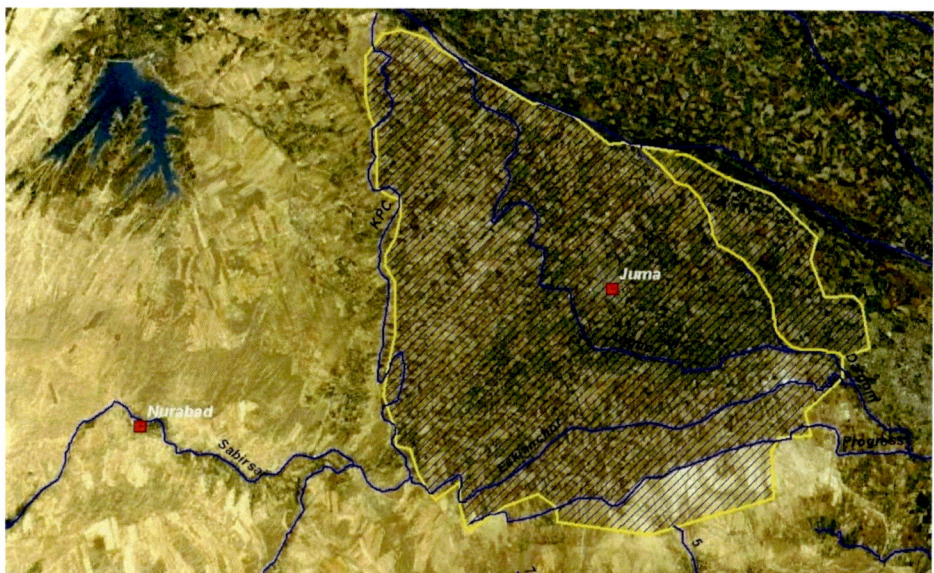

*FIG. 22: GIS elaboration of a satellite image (source: Google Earth TM) representing the two WUAs (Pastdargom in the left and Talligulom Meva Uzum in the right), (two territories delimited by yellow lines) established by the RESP II project in 2009, ( approx. scale: 1:500.000, N↑).*

The WUA staff added that, due to these conditions, they were able to organize meetings among the water users (with large farmers, in particular), three times a year, which were strongly promoted by the RESP II project implementers; furthermore, the director claimed that if the farmers have some issues with water delivery, now it is easier to meet and discuss the problems, thanks to shorter distances from the villages to the WUA headquarters.In addition, their *miraab* spend more time in the fields in comparison with those working for the former Pastdargom WUA. Therefore, although a governance structure characterized by official councils has yet to be created, the participation of the water users, at least in the allocation procedures, seemed to increase. However, it should be emphasized that only the farmers (185, owning on average 25–40 ha) involved with state crops (cotton and wheat) and grapes are involved with the WUA, while the household plot owners are not integrated into the association, but receive the water for irrigation for free. The WUA director, when asked about the ISF, stated that since 2008, as in Urgut and Nurabad WUA, the water charges have been paid by the water users, although with frequent non-compliance; in contrast, since the establishment of the Talligulom Meva Uzum WUA in 2009, the ISF has been strengthened, partly due to the project's support, and nowadays the organization is able to rely, to some extent, on the fees collected from the farmers,[401] As previously mentioned, the RESP II project also focuses on the restructuring of water facilities, and the maintenance of secondary canals. The establishment of new measuring points on the small water courses arising from Siab canal allowed an improvement in water measuring and allocation, which induced the farmers to pay the fees. When asked about the newly established WUA's performance, the majority of the farmers interviewed claimed that the water delivery services — in terms of both the timing and the amount — had significantly improved, and in the last few years water scarcity periods have not occurred; furthermore, the water users stated that nowadays it is much easier to deal with potential issues due to the reduced number of water users in the organization.[402] In addition, since water management and delivery is up to the staff of the former *sovkhoz* Kasymov, they have been on good and close terms with them for a long time. Therefore the evidence in Talligulom Meva Uzum WUA has showen that the RESP II pilot project has led to positive initial outcomes, from organizational perspectives, as confirmed by the staff and the water users.

### *Strengthening the donors rationale Vs the local political initiatives*

Due to the positive outcome of the pilot-project area, in 2010, the international donors, in accordance with the core principles and the aims of the RESP II project, induced the Pastdargom WUA members and the district *hokimyat* to implement further institutional and organizational changes in order to strengthen the IMT and improve the WUAs' performance. The idea of the project partners — the WB, SDC, and SIC — has been to divide the Pastdargom WUA territory, which split and separated from Talligulom Meva Uzum WUA in 2009, into six new water associations according to hydrographic principles (FIG. 24).[403] In the first months of 2010, the donors and the implementing agencies organized several meetings with all the stakeholders (WUAs' members, Dargom ISA's staff, and the farmers) in order to decide the territories and boundaries of the new WUAs; those meetings were integrated with seminars which focused on practices to improve water conservation, and secondary and tertiary canals mainte-

---

[401] Personal communication with Talligulom Meva Uzum WUA's director, Kasimov, May 2012.

[402] Personal communication with Talligulom Meva Uzum WUA farmers, Kasimov, May 2012.

[403] Personal communication with Talligulom Meva Uzum WUA's director, Kasimov, and with SDC' members, Tashkent, May 2012.

nance — and also addressed ways to encourage payment of water fees and the farmers' participation in the WUAs' tasks. Furthermore, regarding the ISF, during these meetings the average fee for water services (16,000–20,000 sum/ha for one year – 6/7 USD) in the new WUAs was decided, according to the average fees of the other organizations of the Zeravshan BISA. Finally, six WUAs within the territory of Pastdargom district were designed and organized (FIG.23):

| Pastdargom WUA | 6370 ha |
|---|---|
| Anghor-Pastdargomsky WUA | 8146 ha |
| Kurilishkok WUA | 6000 ha |
| Konciorbog WUA | 10.000 ha |
| Bakorbogli WUA | 3504 ha |
| Progress WUA | 10.082 ha |

According to the project's rationale, the territories of the new six WUAs were designed *ex-novo* considering the formal hydrographic boundaries, referring partly to the main canals and partly to the secondary ones. Nevertheless, as it was claimed by IWMI experts, it is difficult to understand and verify whether the new WUAs are really based on hydrographic principles. A technical topographic map showing the measuring points and the hydroposts and further data would be necessary to clarify this issue, but unfortunately it was impossible to get.[404] However, regarding the WUAs, the rationale on which the RESP II project is based does not present significant differences in comparison with the IWRM's Fergana Valley project, where WUAs based on hydrographic boundaries have been established in the last years, as the SDC and the SIC were involved in both projects. As previously mentioned, part of the new WUAs' territories refer to the master canals and the others were designed according to the secondary ones: the ones referring to the master water courses are Kurilishkok and Bakorbgli WUA (KPC canal), Anghor-Pastdargomsky WUA (Anghor canal), and Pastdargom WUA (Dargom canal).[405]

---

[404] Personal communication with IWMI members, Tashkent, October 2012.
[405] Personal communication with Pastdargom WUA' sdirector, Juma, April 2012.

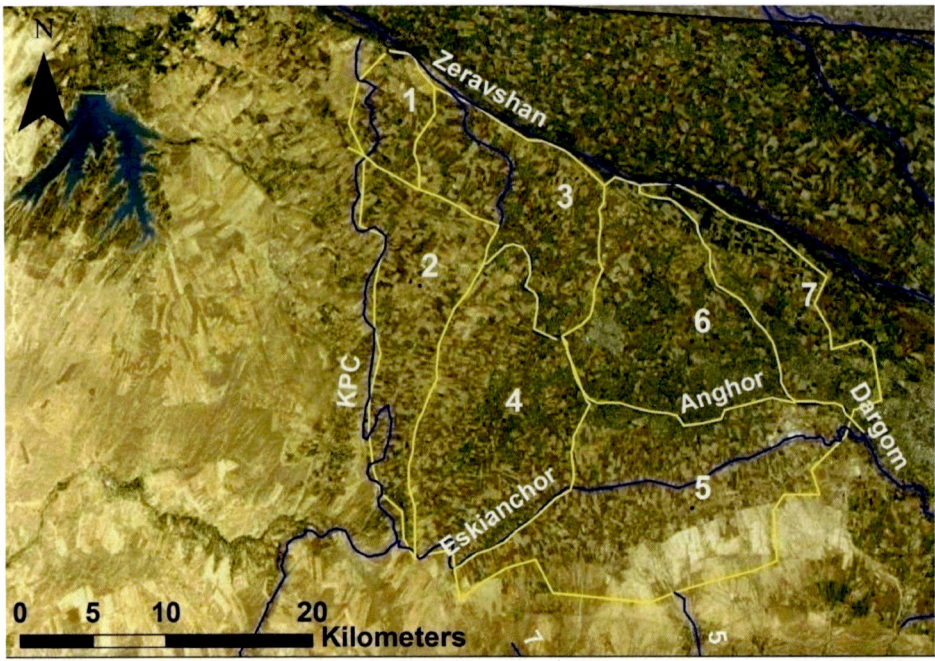

*FIG. 23: GIS elaboration of a satellite image (source: Google Earth TM) representing the territories of the new established WUAs (delimited by the yellow lines) according to the RESP II project: 1, Bakorbogli WUA / 2, Kurilishkok WUA / 3, Anghor-Pastdargom WUA / 4, Konciorbog WUA / 5, Progress WUA / 6, Pastdargom WUA / 7, Talligulom Meva Uzum WUA.*

The WUAs in Pastdargom district are the first organizations in Samarkand province to be based on hydrographic boundaries; in addition, they are also the first WUAs established according to the National Law of 2009, which states that the water users organizations must be designed according to this territorial rationale. The national measure also focuses on the directors, stating that they have to be farmers or water users chosen by the members of the WUAs. The evidence and the data collected have shown that the WUAs' members can suggest one water user as director, but ultimately he has to be approved by the district *hokimyat*; in most of the Pastdargom district WUAs, the heads were finally appointed by the local governors. When the director of Pastdargom WUA was asked about this issue, he claimed that as the new WUAs have only recently been established, they need a director and staff who have already been engaged with both technical and organizational capacities, in order to avoid any problems due to lack of experience; therefore most of the WUAs' heads are the former directors or technicians of *kolchoz* and *shirkat,* as occurred in the Talligulom Meva Uzum WUA, previously analyzed.[406] Due to the uniqueness of Pastdargom district in the water management scenario of Samarkand province, both the WUAs' heads and water users were interviewed in the newly established WUA in order to get an overview of the RESP II project implementation and also to analyze whether the outcomes that emerged in Talligulom Meva Uzum WUA have been widespread in the other organizations too. Obviously the split of Pastdargom WUA into six new associations has led to a significant reduction of the number of water users, which nowadays ranges from 60 (Bakorbogli WUA) to 180 (Progress WUA). Most of the stakeholders agreed that smaller WUAs' territories and the reduced number of farmers have led to relevant benefits to them inthe last years; despite the new organizational structures, the staff and the *miraab* were able to improve water allocation procedures, such as water requests and delivery timing, and when problems emerge, the relations between the farmers and the WUA's staff are close and collaborative.[407] Furthermore, the disputes among the farmers due to the theft of water, which occurred quite often a few years ago, are nowadays reduced. The water facilities' restructuring at the farm level ensured by the international donors, which is going on throughout the district, has led to significant support for improving water allocation practices. In addition, the *miraab* — in particular, the ones who worked in the *shirkat* but have not worked in the former Pastdargom WUA and hence are not used to delivery water to the peasant farmers — were trained by the staff of the implementing agency, SIC. Focusing on the ISF, as mentioned before, during the meeting carried out in 2010 among the WUAs' staffs, a price for water delivery services was decided and in the last couple of years it has been possible to affirm, as claimed by the WUAs' heads, that the number of the water users who have regularly paid strongly increased; a higher fee collection has allowed the WUAs' staffs to pay the *miraab* regularly and, furthermore, to improve the maintenance of the tertiary canals and the connected outlets and measuring points.[408] As stated by the farmers interviewed in Konciorbog and Anghor-Pastdargom WUA, the water users were able to pay more often due to the credits given by the World Bank to make them more responsible regarding water use and to try to limit water waste.[409]

---

[406] Personal communication with Pastdargom WUA' sdirector, Juma, April 2012.

[407] Personal communication with Konciorbog WUA's farmers, Timurhodja, April 2012.

[408] Personal communication with Anghor-Pastdargom and Konciorbog WUA' s heads, Pastdargom district, May 2012.

[409] Personal communication with Anghor-Pastdargom and Konciorbog WUA's farmers, Pastdargom district, May 2012.

172

*FIG. 24: Portion of the Soviet Topographical map (1:100.000, 1976) representing the territory of Anghor-Pastdargom WUA, its settlements, and the secondary and tertiary canals network.*

### The donors rationale led an heterogeneous WUAs context: differences of power and inequities

The context of the WUAs' performance and of the RESP II project changes highlights important differences between the upstream and downstream areas of district (Kurilishkok and Bakorbogli WUAs). In Kurilishkok WUA, both the staff and the farmers interviewed claimed that water scarcity is a regular issue in these lands; as the KPC canal is situated in a peripheral, tailend of the canals system, often the water flow is not sufficient to supply water to all the farmers' lands. Furthermore, the part of the WUA territory supplied by the secondary canals arising from the KPC is still more disadvantaged; another reason which has emerged from the fieldwork is the soil quality, which is worse in comparison with the other lands of the district due to the physical position at the limit of the irrigated area close to the Nurabad steppes.[410] The WUA director claimed that these problems have affected these lands for several years, but he added that the recent changes supported by the RESP II project have not led to significant improvements; for two years, since the WUA was created, they have had to deal with these issues by themselves. Similar issues emerged in Bakorbogli WUA,

---

[410] Personal communication with the Kurilishkok WUA's staff and farmers, Gulistan, May 2012.

which lies in the north-western part of the district at the tail end of the KPC canal and the secondary canals arising from Anghor. Furthermore, this territory was rain-fed livestock land until the 1980s, when two secondary canals were built to convert it into an area oriented towards irrigated agriculture; nevertheless, nowadays, due to these features, the soil is still worse in comparison with the other part of Pastdargom district. Farmers in Cimboi village argued that Bakorbogli WUA often is not able to provide steady and regular water delivery because of the water shortages in the canals; they added that in the neighboring Anghor-Pastdargom WUA the farmers use more water than the allotted amount, without any admonition from Dargom ISA. The WUA staff informed the ISA but no improvements have occurred in the last months.[411] Furthermore, in addition to those natural and management issues, the Bakorbogli WUA relies on a small and less experienced staff regarding water issues, as they come from the former collective farms that were oriented towards livestock farming. The evidence has shown that several disputes regarding water supply and allocation occurred both among the farmers themselves as well as between the farmers and the WUA staff. Some of the farmers often do not receive water according to their contract and the agreed-upon time schedules. Most of Bakorbogli WUA water users, included the household plot owners, agree that in this territory the RESP II project has not yet led to any technical and organizational improvements and benefits, probably because of their peripheral location. With the exception of some meetings organized with the donors in Juma, no members of the implementing agency have come to this territory and no water facilities' restructuring have been carried out.[412] Kurilishkok and Bakorbogli WUAs, due to these analysed issues, are featured by a less developed level of integration and participation compared to the other WUAs. Although it was previously mentioned that in Talligulom Meva Uzum, Pastdargom, and Konciorbog WUAs, the relations between the staff and the water users significantly improved, due to a reduced number of members and to a shorter physical distance between the fields and the WUA offices, it was pointed out that a governance structure, characterized by a participatory approach, has not been created yet. Even though in Talligulom Meva Uzum and Pastdargom WUAs some well-organized meetings with the water users were scheduled, in the other WUAs just informal meetings took

---

[411] Personal communication with Bakorbogli WUA's farmers, Cimboi, May 2012.
[412] Personal communication with Bakorbogli WUA's farmers, Cimboi, May 2012.

place, hence the RESP II project rationale to make the adoption of WUAs' councils a common practice,it is still a long way off from being implemented.

### The RESP II WUA is not an initiative for the Zeravshan basin: the predominance of the local logics

As mentioned by the directors of Pastdargom and Talligulom Meva Uzum WUAs, the implementation of the IMT and of the IWRM's pillars in general is a very challenging process which will require a great deal of time to be fully implemented; according to their opinion, it is already an achievement that the proposed reforms were considered by the local stakeholders, due to the donors' activities, and the organizational and structural changes are in process. In addition, it is also challenging for the farmers to assume more responsible behaviour regarding water use and payment and participate to some degree in the decision-making processes, as they have been used to a top-down approach in these practices.[413] Surely, most of the farmers and the local stakeholders had never been involved with governance structures, such as organized meetings, until a few years ago when the project was designed. According to the donors and to the local experts, the project will last for three years, closing in 2015; the staff of Pastdargom WUA, when asked about the future of water management and the current WUAs' organization in Pastdargom district, maintained that they hope to go on with the framework promoted by the RESP project, strengthening the new institutional and organizational structure and improving its performance, though it will be even more difficult without the organizational and economical support provided by the donors.[414]The evidence has shown that, in considering the Pastdargom district as a whole, the RESP II project has so far led to positiveinitial outcomes; most of the stakeholders interviewed, both water users and WUAs' heads, claimed that the implementation of the project — in particular, the physical and "human" reduction of the water organizations — led to benefit and improvements. Certainly the economic loans to the WUAs and to the farmers have allowed the rehabilitation of part of the farm-level canals and measuring points, leading to better working conditions for the *miraab* and an improved water delivery service to the farmers. Furthermore, the financial availability and a better performing water allocation system have induced the farmers to start paying the water service fees, according to the prices decided by the donors, which is an

---

[413] Personal communication with Talligulom Meva Uzum and Pastdargom WUAs, Pastdargom district, April 2012.
[414] Personal communication with Pastdargom WUA's staff, Juma, April 2012.

essential component for achieving fair WUAs practices and performance. Nevertheless, it is important to emphasize that the project outcomes have not been homogeneous throughout the district; the evidence has shown there is still an upstream-downstream issue, combined with a more evident centre-peripheral one, as highlighted in Kurishilkok and Bakorbogli WUAs. Those territories, due to their physical characteristics and probably to less influential political power, from both district-level authorities and international donors, have not benefited from the project to the degree that the centre area has in the last years. Therefore, what finally emerges is that in the Zeravshan basin scenario, an international project promoting the main framework, which should havebeen supported by the government according to the enacted measures, has been able to establish a different WUA context, compared to the other districts included in Zeravshan BISA. The Uzbek government has allowed the international organizations, through the project, to support and start implementing, although with several challenges, a new perspective of local water management based on the international rationale, quite far from that local one that emerged just outside the project. In the next years, when the financial support to the district's stakeholders will cease, the WUAs should be able to rely on the members' funds and capacities. What emerged from the national and province authorities' policies is an ambiguous and questionable behavior towards both the development agencies and the IMT and IWRM framework: on the one hand,in the last years, the government has nationally allowed the international organizations to set-up different projects in certain districts to spread the international wisdom about water management. On the other hand, in all the regions not involved in the blueprints, the main authorities have kept the water management status quo without any organizational and institutional changes. Furthermore, the last national measures — for instance, the 2009 measure on WUAs based on hydrographic boundaries, supported by the donors and enacted by the government — are still far from an effective implementation. This scenario clearly emerges in the Middle Zeravshan valley: the Samarkand province government, in connection with the national government and the other basin-level stakeholders, has allowed the Pastdargom district to join the WB and SDC RESP Project as a pilot area in order to be legitimized in the other districts to keep enable the local deviance from the mainstream water management structure. This strategy can be understood as a way to get an international reputation for fairness, collaborating with international actors and consequently receiving loans and financial aid, while at the same time keeping strong state control in water and agricultural processes characterized by state quotas and top-down approaches. One question that has arisen concerning the outcomes under discussion has been whether the Pastdargom district example could be extended to the other administrative units of the Zeravshan BISA; the evidence has shown that since 2009 no ideas or measures have been undertaken either by BISA's governing board or the WUAs' heads to change the organizational structure of the district water users associations. Therefore, it seems that the basin and local level water bureaucracies will oppose the widespread adoption of the project rationale in the other districts, in order to preserve the current political-economic status quo.

# 6. THE LOGICS OF THE BASIN / LOCAL LEVEL WATER REFORMS IN KAZAKHSTAN: THE ARYS VALLEY

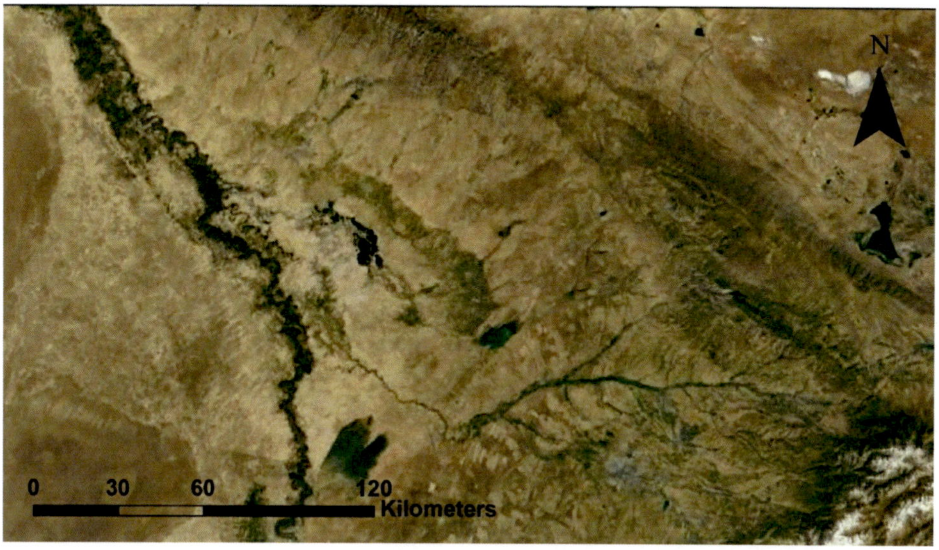

*FIG. 25: Satellite image (source: Google Earth TM) of the Arys Valley.*

## 6.1 A GEOGRAPHICAL OVERVIEW OF THE VALLEY

The Arys valley, as it was pointed out in Chapter 2, was selected as a Kazakh case study for multiple reasons, ranging from territorial to economic-political ones. The valley, which lies in the southern-central part of Kazakhstan, is one of the largest and most important irrigated areas of the country. Contrary to Uzbekistan, which is characterized by several irrigated areas lying throughout the republic, in Kazakhstan all of the irrigated areas are located in the south, mostly close to the border mountain ranges (FIG.25). Due to the climatic and environmental features, in northern and eastern Kazakhstan, it is possible to conduct agriculture without irrigation, while in the southern branch of the country, irrigation is essential and strategic, in particular in the basin's downstream areas.[415] Contrary to the Middle Zeravshan valley, whose territorial development dates

---

[415] POMFRET, R., 2007. "cit.".

178

back to ancient times, the Arys valley — due to the Kazakh historical and social context featured by nomadic pastoralism — has been affected by hydraulic structural changes since the 1950s, according to the Soviet *hydraulic mission.*[416] The Arys river is the most important tributary of Syr-Darja in Kazakhstan's territory, flowing in N-W/S-E direction at a latitude of 42° N; the catchment of the river, which is considerably smaller in comparison with Zeravshan's, has an area of approximately 15,000 km3 and can be divided into two parts — the upper-middle river valley featured by the mountains and the foothills and the lower valley featured by the plains and the steppes. Arys river originates in the Talas-Alatau mountain range, which is included in Western Tian-Shan mountains, after the merging of several small streams, with an average altitude of 2000 m. a.s.l. (FIG.26). Running a total of 378 km in length, the Arys river flows into the Syr-Darja and the average run-off is 46 M3/s.[417] Furthermore, in the upstream-central valley, Arys is supplied by the flows of several tributaries — Boraldai, Aksu, Badam, and Mashat are the main ones. Therefore the river, due to its physical and climatic conditions, is characterized by a simple regime with two hydrological seasons.

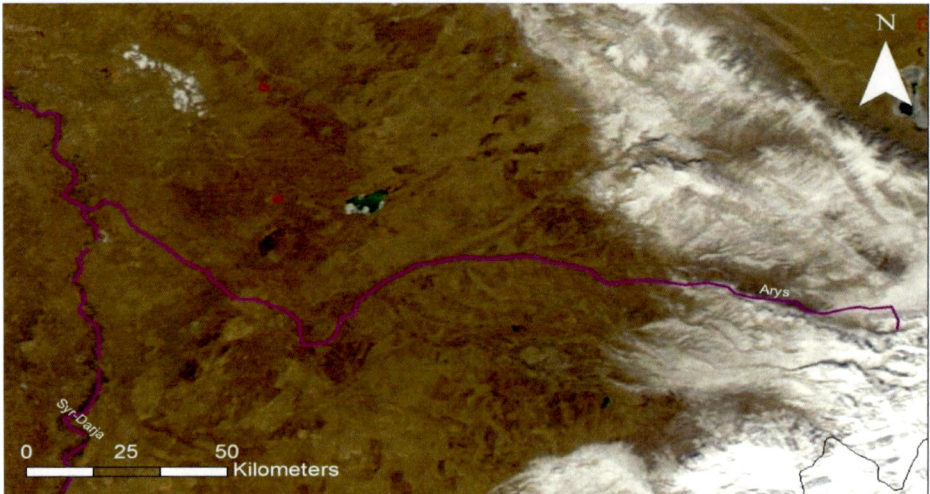

*FIG. 26: GIS elaboration of a satellite image (NASA-Modis,2003), representing the Arys river and its valley.*

---

[416] Personal communication with SouthKazahstan Province Water Department's members, Shymkent, November 2011.

[417] Personal communication with Aral-Syr-Darja Basin Agency' members, Shymkent, November 2011.

The flood season is divided between the snow melt in the spring — the highest due to the relatively low altitude of the springs — and the ice melt in the summer, featured by the streams which arise from the Talas-Alatau glaciers. The second hydrological season is in winter, when, due to the low temperatures, ice and snow cover, the flow is considerably low. Administratively the whole Arys valley is included in the South-Kazakhstan province; this territory includes most of the Central Asian physical-environmental features, such as irrigated plains (less wide in comparison with Zeravshan), steppes, foothill areas, and mountains, as previously pointed out in the discussion of Zeravshan valley. The irrigated area stretches a few kilometres in the upstream and central part of the valley, lying close to the river branches and due to the absence of extended irrigation schemes; this section is surrounded by rain-fed mountains and sloping hills (Talas-Alatau range in the S and Karatau range in the N). The difference in altitude in this part of the valley is not particularly significant, ranging from 800 meters a.s.l. close to the springs (the main one is 1200 metres a.s.l.) to 300 a.s.l.on the lower side. In contrast, on the valley's downstream side, the irrigated areas are wider due to the presence of the irrigation schemes, which will be further analyzed; these irrigated areas are located along the river and the main canal, surrounded by the steppes. Regarding the local annual rainfall, it differs significantly according to the proximity to the mountains, as in the Zeravshan valley; the approximate range is from 150–200 mm/y in the downstream part of the valley close to the Arys's confluence into the Syr-Darja, then 400 mm/y in Shymkent, and up to 450–500 mm/a on the upstream side (Tyulkibas district).[418]

### 6.1.1 Transforming the territory: the hydraulic mission

Focusing on the Arys valley's territorial features, in contrast to Zeravshan valley, only the central-lower part can be considered a hydraulic territory since the higher one has not been affected by the design of the canal networks. Therefore, the central-downstream section has been the part featured by hydraulic infrastructure construction and subsequent territory transformations carried out during the Soviet Union beginning in the 1950s. As mentioned above, it is divided into two different and separate parts: the Arys-Turkestan irrigated area and the Shaulder area. In Arys's mid-stream right bank, not far from Temirlan village, Arys canal, designed at the end of the 1940s, arises from the river and after 25

---

[418] Data collected in the Department of Geography, Auezova State University of Shymkent, November 2011.

km flows into the Bogun reservoir. In the first 10 km of its stream, it crosses the steppes, while in the second part, the canal irrigates a branch on its western bank. Close to the off take of Arys, another canal arises from the river, Karaspan, flowing in a south-eastern direction for 26 km and increasing the irrigated area located on the river banks. At the end of Arys canal, the Bogun reservoir was built for water storage (max. capacity 370 million m3.) during the wintertime and for releasing water at the beginning of the cropping season. From this huge infrastructure — the biggest of the canals system — arises the Arys-Turkenstan canal (ATK), built during the 1950s and featured by an average flow rate of 40 M3/s; the canal has a total length of 92 km and irrigates a total of 55.000 ha through an irrigation scheme of 55 secondary canals (average length 10–15 km) arising from the ATK outlets (FIG.27).[419]

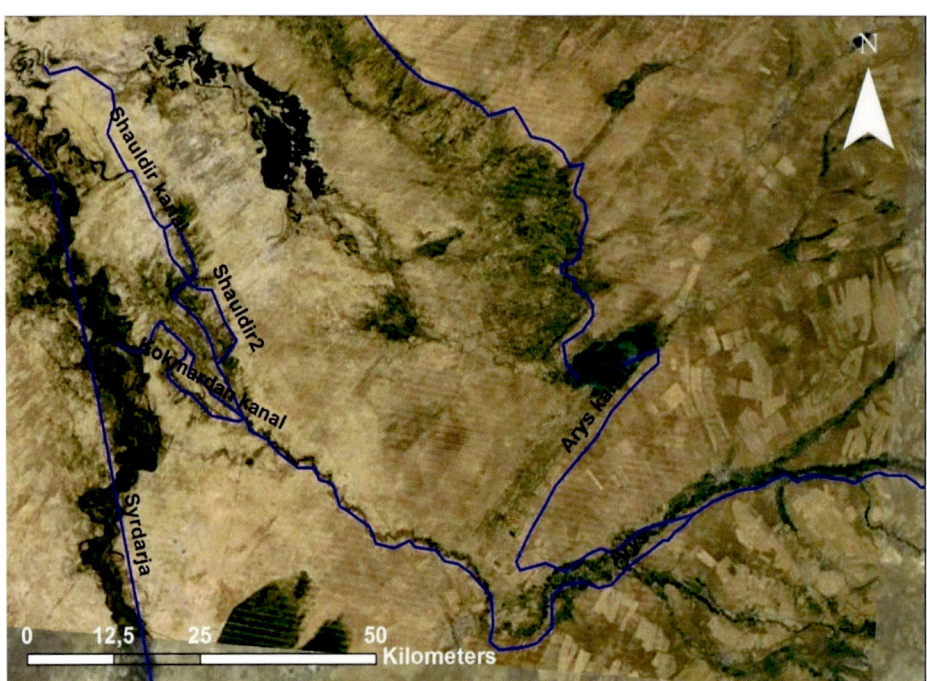

*FIG. 27: GIS elaboration of a satellite image (source: Google Earth TM) representing the canals' system (blue lines) and the irrigated areas (green zones) of the central-downstream side of Arys valley.*

---

[419] Personal communication with South-Kazakhstan province water department's members, Shymkent, October 2011.

The second irrigation scheme built during the Soviet Union (1960–1970) lies 25 km before the Arys's confluence into the Syr-Darja; from the small Shaulder barrage, arises a canal network featured by three main canals — Kokmardan, Altimbekov, and Shaulder — featured by an average length of 15–20 km and a low run-off (5–8 M3/s).[420] Shaulder's irrigated area ranges approximately 35.000 ha, but in the last two decades, due to frequent water shortages and soil salinization, only 20 to 25.000 ha have been used for irrigated agriculture.[421] Focusing on the main crops farmed in the valley, wheat, fodder, cotton and cor-nare the most widespread, along with the fruit and vegetables that are cultivated in households plots. Most of the cotton farming of Arys valley lies in the irri-gated area supplied by the ATK canal; this is the northernmost region for cotton farming in the world (42° N). As in the Zeravshan valley, the total irrigated area of Arys valley has not increased since the collapse of the Soviet Union; no ca-nals have been built in the last two decades. Nowadays the irrigated area of the entire Arys valley is in the range of 240.000 hectares, but due to soil salinization issues, in particular in the downstream part, only 160.000 to 170.000 are effec-tively used.[422] A new water infrastructure, one of the few designed in the Central Asian region since 1991 — strategic for water storage but not relevant for irriga-tion — was built in 2010, close to the Syr-Darja river, 80 km northern Chardara reservoir, and named Koksarai *kontroregulator*. This huge reservoir's main pur-pose is to collect and save Syr-Darja winter flood waters and then release water for irrigation during the vegetation period; in addition, it also works to prevent flooding in the Syr-Darja valley.

## 6.2 MANAGING WATER AT THE BASIN LEVEL: DISPUTES OF SCALE AND AUTHORITIES

### 6.2.1 The regulation of the conflicting basin- level organizational structure

In Arys valley, according to the water reforms highlighted at national level in Chapter 4, nowadays the water management organizational structure is charac-terized by two main authorities: the Aral Syr-Darja river basin organization (*BWO*) and the South-Kazakhstan Republican state enterprise *(RGP)*. Whereas,

---

[420] Personal communication with South-Kazakhstan province water department's members, Shymkent, October 2011.

[421] Personal communication with Otrar district water department's director, Shaulder, May 2012.

[422] Data collected in South-Kazakhstan RGP, Shymkent, November 2011.

as it was pointed out in the previous chapter, the Uzbek scenario of implementation of national reforms at the basin level has been heterogeneous, showing different trajectories due to the province and local political powers, in Kazakhstan those processes have been more regulated and homogeneous. According to the Water Code enacted in 1993, since the collapse of the Soviet Union, water management in Arys valley was characterized by the Aral Syr-Darja *BWO*, a branch of the Committee of Water Resources, based on hydrographic boundaries, and by the South-Kazakhstan province water department, funded by the province's budget and based on administrative boundaries. As previously pointed out, in the first years after independence, although the separate tasks of the two authorities seemed clear and formalized, the organizational scenario and the management tasks were quite fuzzy and indistinct throughout Kazakhstan. In the South-Kazakhstan province, according to the data collected, vague disputes regarding tasks and responsibilities among the two authorities emerged. Specifically, due to the institutional and organizational crisis within the agricultural and water sectors after independence, the Aral Syr-Darja BWO, funded by the Committee of Water Resources, suffered organizational lacks and financial shortages to a greater extent in comparison with the South-Kazakhstan Province Water Department.[423] Therefore, as it was previously debated, the tasks and responsibilities of the two authorities, the controlling body and the technical one, were regulated in 1996 through the Decree on the Differentiation of Functions between the River Basin Agencies and the Province Water Departments. Nevertheless, although this measure had to balance the different functions, the province water departments gained a more influential role and a financial increase since they were transferred to the state's central administration and renamed as Republican State Enterprises (*RGP*).[424] Therefore, the South-Kazakhstan province water department changed into the South-Kazakhstan Republican State Enterprise "*Iujvodkhoz*". This decree formalized the current institutional and organizational structure of water management in Arys valley.

---

[423] Personal communication with Aral Syr-Darja BWO' smembers, Shymkent, November 2011.
[424] ZIMINA, L., 2003. "cit.".

## 6.2.2 A weak government support to the hydrographic principles - despite the formalization of the IWRM

The Aral Syr-Darja River Basin Agency, funded during the Soviet Union as a branch of the Soviet Ministry of Water Resources, was transferred to the control of the Committee of Water Resources (included in the Ministry of Agriculture since 1997) after independence and formalized through the Water Code of 1993. This measure assigned the role of primary water management agency to the *BWO* in its specific territorial jurisdiction. As it is based on a hydrographic unit, according to the Water Code, the Aral Syr-Darja BWO territorially includes the Kazakh section of the Syr-Darja valley and its tributaries, from the Chardara reservoir close to the Uzbek boarder in the south to the Aral Sea in the north; no changes in boarders have occurred in the last two decades. From an administrative perspective, Aral Syr-Darja *BWO* includes two entities, South-Kazakhstan province in the south-east and Kizylorda in the north-west; therefore, in managing water resources, the agency deals with South-Kazakhstan and Kizilorda *RGP*s. Since the river basin agency is a branch of the Committee of Water Resources based in Astana and is a controlling body, it annually receives an allotted amount of water, measured in cubic km, from the central headquarters which is then divided among the two republican state enterprises according to their water needs and irrigation plans. According to the data collected, the total amount of water allocated to the South-Kazakhstan RGP ranges from 3.8 to 4.2 million M3.[425] Therefore, the river basin agency's main responsibilities are the monitoring of water resources and its consumption  as well as the overseeing of water quality and pollution levels of Syr-Darja river, its tributaries and catchment streams; those tasks are not only related to water for irrigated agriculture, but also for urban and village water supply.  Furthermore, although it does not have a technical responsibility, the agency monitors inter-province reservoirs , such as Chardara, which regulates the Syr-Darja's flow and the Aral Sea; due to the importance of its issues in the last years, Aral Syr-Darja BVO is working in collaboration with the Ministry of Emergency and with the inter-republican ICWC. According to the evidence, although the differentiation of functions between the two water bodies was enacted in 1996, until the 2000s the river basin agency suffered organizational lacks and in particular financial shortages, due to the redesign of the official structure of the Committee of Water Resources at the national level. Since the new Water Code was issued in 2003, establishing the

---

[425] Personal communication with Aral Syr-Darja BWO' members, Shymkent, November 2011.

legal framework for the IWRM, the Aral Syr-Darja BWO has gained more support both from the government and the international organizations, in order to
strengthen water management based on hydrographic principles. Furthermore,
the following year the international project "National IWRM and Efficiency
Plan for Kazakhstan" sponsored by the UNDP and the GWP, promoted changes
to the organizational structure of Aral-Syr-Darja BWO, supporting the establishment of bottom-up governance and its implementation through the creation
of the basin councils, according to the IWRM pillars. Nevertheless, as stated by
the interviewed members of the Aral Syrdarja basin agency, the issue of the water code and the related support to strengthen the river basin agencies have not
led to significant changes. On the one hand, the authority was reinforced, in particular from an institutional/political perspective; on the other hand, their tasks
and the organizational features have not significantly improved in the last
years.[426] For instance, the international organizations' members organized some
seminars to develop the concepts of the IWRM, managing water according to
basin principles and an environmental approach, but it has been hard to effectively implement what they announced in the last years; in addition, financial
resources have not significantly improved.

### 6.2.3 A weak support to a basin level participatory water governance

Concerning the establishment of the governance structure, as previously pointed
out, the first River Basin Councils (RBCs) in the Kazakh river basin agencies
were set up beginning in 2006. The Aral Syr-Darja basin council was created at
the end of 2007 with the financial and organizational aid of the UNDP and the
Committee of Water Resources. Since 2008 the meetings have been scheduled
twice a year at the beginning and at the end of the cropping season. According to
experts in the Hydro-Melioration State Enterprise, which collaborates with the
UNDP, the basin councils include members involved in water management at
different levels (basin, district and villages), such as employees of the Aral Syr-
Darja BWO, the South Kazakhstan RGP, the province and district water departments, the WUAs, and farmers; in addition, occasionally members of NGO
working on ecological issues have been involved.[427] In order to facilitate the
councils' organization, due to the extensive area of the basin and to strengthen
the councils — in particular, concerning different issues besides those organized

---

[426] Personal communication with Aral Syr-Darja River Basin's members, Shymkent, November 2012.
[427] Personal communication with Hydro-melioration State Enterprise, Shymkent, November 2012.

for the whole basin — six different sub-basin councils were designed, mainly according to the inner catchments of the Syr-Darja tributaries. In Arys valley one sub-council was designed for the whole catchment, and throughout the South-Kazakhstan province other governance structures were created in the Makhtaral irrigation system (connected to the one in Uzbek Hungry Steppe) and in Keles river basin.[428] Nevertheless, although these councils can be considered a first step towards a participated and integrated approach in water resources management, lack of organization, interest, and involvement emerged from the water users. Moreover, in the last three years, any kind of elections of the governing board has been effectively supported (both in the basin council and sub-basin); in Aral Syrdarja Basin Council the head is the director of the agency, who also appoints the head of the of the sub-basin councils. Therefore, the evidence shows that the bottom-up practices, strongly promoted by the donors and the backbone of the project, are still immature and undeveloped; that is, the councils' organization is still oriented towards a top-down approach. From the interviews conducted with the water users and the WUAs' staffin the villages of Arys valley, lack of interest emerged regarding the river sub-basin councils; most of the interviewees were not aware of these new organizations, including the WUAs directors and the members of the district water departments. Among those who participated, both in the basin and sub-basin councils, they claimed that those councils are not necessary to improve water management practices and it would be more important to strengthen the province/district level departments which deal with irrigation systems' operation and maintenance; furthermore, they stated that the WUAs councils, even though quite rarely involved, are more useful in dealing with irrigation issues.[429] Therefore, the evidence has clearly shown that, although the councils have been sponsored and established, the importance of a participatory approach at the basin and sub-basin level has not really convinced or involved the water users at a local level.

---

[428] Personal communication with Hydro-melioration State Enterprise, Shymkent, November 2012.

[429] Personal communication with WUAs' members and farmers in Arys valley, April and November 2012.

### 6.2.4 A move towards recentralization? The technical body, from province to state budget

Focusing on the other actors involved in water management at the province/basin level, the South-Kazakhstan Republican State Enterprise was established during the Soviet Union as the South-Kazakhstan province water department, funded and control by the province authority; this status was maintained until 1996. As widely pointed out above, the province water authority was renamed "*Iujvodkhoz*-South-Kazakhstan RGP" and its institutional structure was changed through the issue of the decree on the "Differentiation of functions between the river basin agencies and the province water departments"; hence, since 1996 it has been funded through the republic's budget and is no longer within the province's budget. South-Kazakhstan RGP *Iujvodkhoz* is territorially based on the administrative principle, referring to the province's boundaries and no changes regarding territory have occurred since either the collapse of the Soviet Union or the 1996 decree. Concerning the water resources, the water authority's territory is crossed by the Syr-Darja, flowing from the Chardara reservoir to Kizylorda province, including the whole Arys valley and its irrigation system, the Keles valley, and the southern branch of the Machtaral irrigation system. The total irrigated area under the South-Kazakhstan RGP's control reaches 525.000 hectares, while the rest of the territory is featured by the steppes in the west and the rain-fed pastures and mountains in the east and in the north. Nevertheless, according to the interviewed members of the authority, nowadays the irrigated land measures approximately 420.000 hectares, due to the unavailability of water and to land salinization issues, in particular in the downstream areas.[430] Considering the whole irrigated land (525.000 ha), 240.000 ha are supplied by Arys river and its tributaries, but due to the issues mentioned above, only 170,000 ha are used for irrigated agriculture nowadays. As previously analyzed, South-Kazakhstan RGP, according to the 1996 decree, is a technical body, dealing with water facilities' operation and maintenance; hence its main responsibilities are the operation of water system units, the maintenance and improvements of technical infrastructures conditions, and the water supply to the authorities and organizations at the local level (district water departments and the WUAs). Regarding the water facilities' operation and maintenance, South-Kazakhstan RGP is responsible for the primary level canals and reservoirs: Arys canal, Arys-Turkestan canal, Karaspan and Kulan canal (Ordabasy district), and

---

[430] Personal communication with SouthKazakhstan RGP' members, Shymkent, November 2011.

Altymbekov, Shauldir, and Kokmardan canals (Otrar district). According to the water authority experts and the Ministry of Emergencies, in a few years the Koksarai reservoir and connected canals system (*Koksariskyi Kontroregulator,* built in 2011) will also shift from Ministry supervision to RGP's.[431] Concerning the water allocation practices, once the authority receives the total water allotment from the Aral Syrdarja River Basin Agency, it divides it among the district water departments and the WUAs, according to their water needs and the irrigation plans which have to be prepared and formalized with the River Basin Agency at the beginning of the cropping season. Technically, the primary canals' water flow is controlled and divided by the hydro technicians and *miraab* through the outlets into the secondary canal network. Besides the district water departments and the WUAs, the South-Kazakhstan RGP staff also allocates water to the farmers' lands which lie close to the primary canals; according to the authority's staff, those farmers are in a privileged position in respect to the others, since they do not have to deal with district level water authorities and connected potential issues.[432] Due to the relevancy of the Bogun reservoir for the central-downstream Arys valley irrigation system, SouthKazakhstan RGP has a technical branch in Bogun village; this department is responsible for the Bogun reservoir and its water release (16 employees) and for the Arys-Turkestan canal and its 22 outlets (14 employees). The RGP's Bogun branch has also profited in the last decade from the institutional shift from the province budget to the republican state budget. According to members of the South-Kazakhstan RGP, the shift from the province water department to the republican state enterprise has benefited the water authority and to all the water users, in terms of rehabilitation and improvements to the province's irrigation system. When asked about potential disputes or ambiguous relations among their authority and the Aral Syr-darja River Basin Agency, they claimed that since 1996 their tasks have been officially separate and they are collaborating together for widespread improvement of water resources management in Aral Syr-darja basin. In addition, RGP members claimed that even before 1996 disputes between the two authorities did not occur.[433] Furthermore, the hydro technicians of the Bogun RGP branch claimed that starting from the 2000s the river basin agency has received significant insti-

---

[431] Personal communication with South-Kazakhstan RGP's members and expert of the Ministry of Emergencies, Koksarai reservoir, April 2012.

[432] Personal communication with South-Kazakhstan RGP's hydro technicians, Bogun, November 2012.

[433] Personal communication with South-Kazakhstan RGP' members, Shymkent, April 2012.

tutional and financial support from the donors-based projects, which also allowed the establishment of the basin council and local sub-councils; hence it was assumed that the state had to support the RGP *Iujvodkhoz* with an increase of finances.[434]

### 6.2.5 Strengthening the leading role of the RGP at the basin level

Nevertheless, in the last two to three years, since 2010, the water authority has received a considerable increase of financial resources which has allowed the restoration of some stretches of the Arys-Turkestan canal and related outlets and secondary canals; also, with funds from the province's budget, part of the Otrar district's primary level irrigation system has been renovated.[435] Besides these technical restorations, evidence has shown that according to the institutional and organizational issues that have occurred at the district level involving the district water departments and the WUAs (which will be analysed in depth in the next sections), the South-Kazakhstan RGP is going to gain a leading role in water resources management at the district level as well; this may be possible with the water authority's potential control of the secondary canals level and water allocation directly to the water users. Therefore, although no kind of real competition or disputes between the two authorities have clearly emerged, in the last years, the evidence has shown stronger state financial and institutional support to the RGP with respect to the River Basin Agency. This political decision seems to be in contrast with the water policies supported by the government since the 2000s, oriented towards the implementation of the IMT/IWRM framework — such as as the new Water Code enacted in 2003 or the participation in the donors-based projects also oriented towards the support of the international water paradigms. Furthermore, this trend can be read as conflictual since the government is more strongly supporting an authority based on administrative principles rather than one characterized by user participation and bottom-up decision-making processes, yet meanwhile, is involved in the recently started (2010–2025) second phase of the "IWRM and Efficiently Plan for Kazakhstan". According to some experts who were asked about those recent measures, this move

---

[434] Personal communication with South-Kazakhstan RGP's hydro technicians, Bogun, November 2012.

[434] Personal communication with South-Kazakhstan RGP's members and with Ordabasy and Otrar district water departments' staff, Temirlan, Shaulder and Shymkent, April 2012.

can be considered as a partial recentralization process of water management at the basin/district level, due to challenges and problems that occurred in the implementation of the IWRM pillars, recognized also by the donors, and a more relevant state involvement in the water issues at the local level.

## 6.3 IMPLEMENTING THE IWRM AT THE LOCAL LEVEL: EVIDENCE FROM THE DISTRICTS

Analyzed the water management context in Arys valley at the basin level, in this paragraph the focus is on the three districts, Tyulkibas, Ordabasy, and Otrar, selected for field research to highlight and point out the water context, the institutional reforms towards the IWRM/IMT, and the related issues at the local level (FIG.28). As mentioned in Chapter 2, the administrative units were chosen according to territorial features as well as their physical position in relation to the Arys river and its irrigation system. Tyulkibas district lies in the upstream part of the valley and is characterized by the irrigated areas supplied by the river and by other small streams, while Ordabasy unit is located in the central-downstream section of the valley and crossed by the Arys and Arys-Turkestan canals. Otrar district lies close to the downstream section of Arys valley and its territory, mostly characterized by the steppes and supplied by the Shaulder irrigation system. Furthermore, throughout the districts, some villages — in particular where the WUAs are based — were selected to conduct fieldwork in order to highlight and understand the institutional and organizational framework for water resources and the local issues. As previously discussed and debated, the reorganization of the water sector at the local level in Kazakhstan —according to the IMT processes and the main framework of the IWRM program in general — has occurred since the end of the 1990s and was formalized through the new Water Code and the Law on WUAs, both enacted in 2003. Therefore these processes will be outlined and debated in selected districts of the Arys valley, focusing on the district water departments and WUAs' institutional, organizational, and operational framework. The analysis will be integrated with data collected through interviews and informal talks with the farmers regarding the authorities' and associations' performances, the local water management lacks and issues, and the potential disputes among the water users and the district-level water institutions.

190

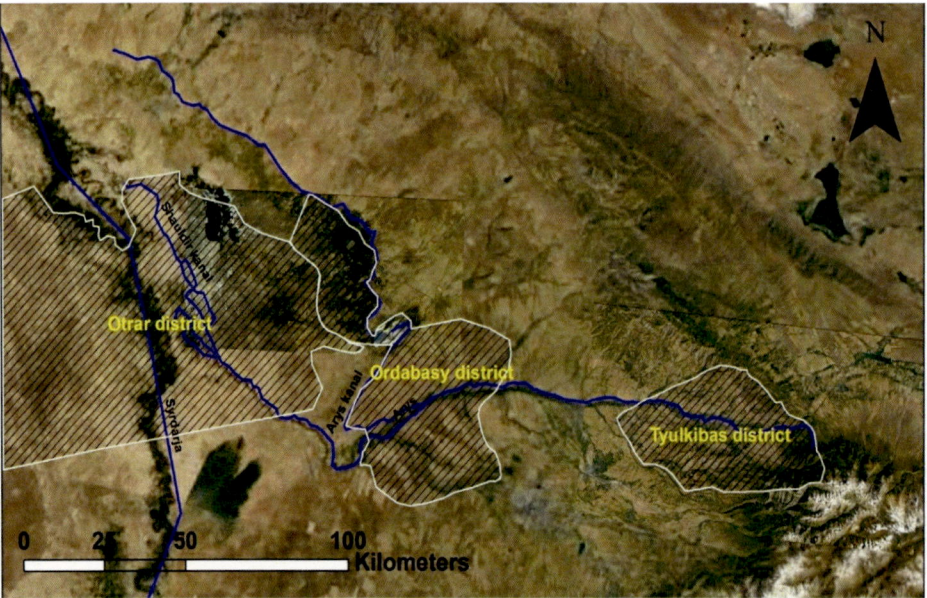

*FIG. 28: GIS elaboration of a satellite image (source: Google Earth TM) representing the three district of the Arys Valley (white shapes) selected for the field-research; in blue the river and the canals network.*

### 6.3.1 TYULKIBAS DISTRICT

Tyulkibas district is entirely located in the upstream part of the Arys valley, in the eastern part of the South-Kazakhstan province, 60 km north-east of Shymkent. The river, fed by the water of several mountain streams, flows in E-W direction at an altitude ranging from 1300 to 650 m a.s.l. The valley is surrounded in the south-eastern part by the western Tian-Shan mountains (Ugam-Talas ranges), which ranges 4050 m a.s.l.., and in the northern part by eastern Karatau mountains, relatively lower, reaching 1700 m a.s.l.; this range divided the Syrdarja catchment in the south from the Chu one in the north[436]. The district irrigated area, mostly located in the centre of the valley close to Arys river' branches, reaches 17.800 ha while the total agricultural one, mainly featured by rainfed land, 64.000 ha; this land lye in the valley' flat areas far from the river and in the hills' slopes. Due to favourable climatic and soils conditions and to water availability, Tyulkibas district has never been affected by the Soviet *hydraulic mission*, even in the more active period, 1960-1980, of territory' trans-

---

[436] SUSLOV, S.P., 1961. "cit.".

formations through water resources management (FIG.29). Therefore it is possible to state that the district is not charactrizeded by properly irrigation schemes, differently from the central-downstream ones, but by a natural network of streams, which part of them where lined and equipped with outlets to supply the fields through tertiary level small canals.[437]

### Establishing the WUA: centralized procedures and uncertainty

According to the evidence and the data collected, the Tyulkibas district water management context has appeared quite anomalous and unstable, when compared to the measures enacted in the last ten years. Despite the national Water Code issued in 2003 supporting the establishment of the WUAs (*SPKV*) and their formalization, in the Tyulkibas district, water management and allocation were under the control of the district water department (*kommunalniy vodkhoz*) based on administrative boundaries and funded with the district budget. In addition, most its staff has not been affected by any changes since the collapse of the Soviet Union. Some water users interviewed in one agricultural cooperative close to Vanovka (the district's chief town) stated that they knew about the WUAs' establishment in Kazakhstan, but the governing board at the district government had not supported this measure since 2003.[438] In 2010 the Tyulkibas district water department filed bankruptcy because of several financial problems: a major decrease in the district budget (already reduced when the state stopped directly financing the district water departments), extremely low revenues collected from the water users and lack of technical staff to control and maintain the water delivery network (most of the members who worked during the Soviet Union retired in the last years). This context totally fits with the analysis of the current national condition of the district water departments, pointed out in Chapter 4. The authority was partly reorganized in February 2011, changed its official status, and shifted into a water users association. Tyulkibas WUA was registered, by the former head of the previous authority, in the district justice department and, once it obtained the authorization, it leased the permission to control and maintain the streams and small canal network and related facilities from the district government (*Rainnovo Akimyat*).[439] Despite the institutional shift, no major changes occurred in either the organization or staff; the

---

[437] Personal communication with South-Kazakhstan RGP's members, Shymkent, April 2012.
[438] Personal communication with one PK' farmers, Tyulkibas district, April 2012.
[439] Personal communication with Tyulkibas WUA' head, Vanovka, November 2011.

director and also the other members (the accountant and hydro technicians) are the same as they were under the previous authority.

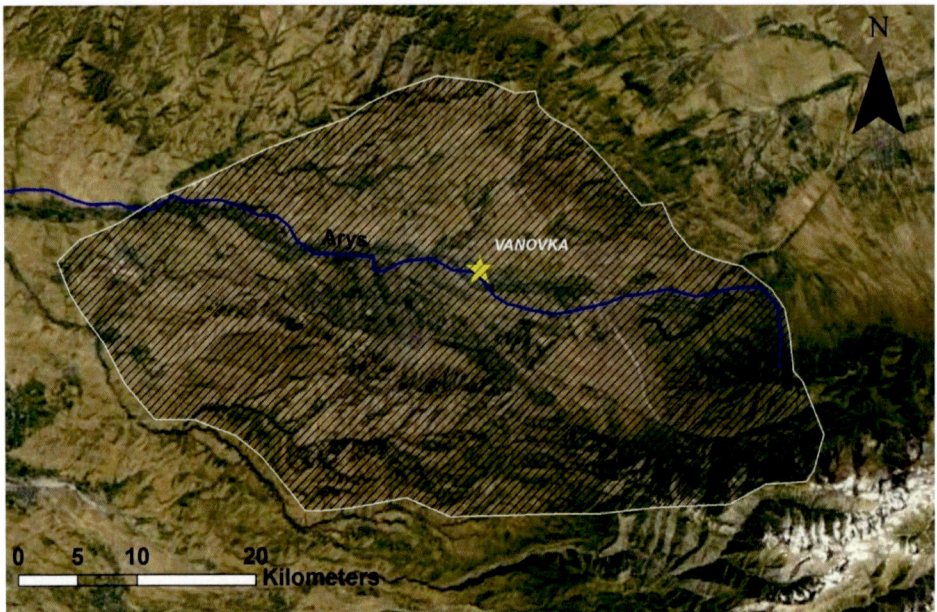

*FIG. 29: GIS elaboration of a satellite image (source: Google Earth TM) of Tyulkibas district (white shape); it is notable the irrigated area close to the river flow and the hilly rainfed area.*

The main difference is in the source of financial support, which no longer comes from the district department, but now must come from the water users. Furthermore, no changes occurred in boundaries, which were kept the same as the administrative ones, despite the IWRM/IMT support to shift the WUAs to hydrographic boundaries. Nevertheless, although the former authority was turned into a WUA, no involvement or participation of the water users in the decision-making processes occurred. According to the Tyulkibas WUA's director, the staff met the water users at the beginning of the 2011 cropping season to sign the contract with them, but no other meetings to discuss water issues have been planned; he added that it is costly and time consuming to organize meetings which are not requested both by the members and the district government.[440] The water users did not mention a lack of participation, but they vehemently

---

[440] Personal communication with Tyulkibas WUA' director, Vanovka, November 2011.

stressed their concern about an increase of the irrigation fee due to the shift from the district water department to the water users association.

### The bankrupcy of the WUA: a crisis of the IWRM / IMT rationale

The WUA worked properly in the cropping season of 2011 although it was still affected by financial shortages mainly due to the lack of payment from the water users. The farmers claimed that there is no sense to pay fees (which increased to 220 tenge/1000 M3 – 1.4 $, instead of 180 tenge, after the WUA establishment) to an organization which is not able to provide streams and outlets' maintenance and fair water allocation.[441] According to the Tyulkibas WUA's director, during the cropping season of 2012 most of the *miraab* could not be paid due to money shortages (only 60% of the total fees due were effectively collected) and for this reason the association members were not able to clean and maintain the water facilities. Although the contract (water facilities leasing) with the Tyulkibas district department was supposed to be in effect until 2013, in October 2012 the WUA decided to stop working. Besides the financial issues — that is, no economical support from the government as well as WUA budget shortages — the director stated that they suffered a severe lack of technical specialists and equipment and they were not able to create a common work ethic.[442] Therefore at the end of 2012, the WUA's members were preparing all the documents for disbanding the association, to certify its failure and arrange for the return of the water facilities control to the district government. The failure of the WUA experience, after only one year and a half, led to an institutional and organizational void in Tyulkibas district's water management. The district court will decide the future administration for control of the water system: if no new WUA is established by the farmers, the facilities' management would be returned to the district government. According to the WUA director, one strategy could be to re-establish the district water department, despite the same financial, technical, and organizational issues which led to the organization's bankruptcy in 2011; the district budget is not enough to cover management costs.[443] Furthermore, the evidence has shown that hardly a new WUA or more smaller WUAs will be created; this new model of water management at the farm level, formalized ten years ago, has not been properly understood by the water users, who have not supported it either financially or with their personal involvement in the decision-

---

[441] Personal communication with the water users, Tyulkibas district, November 2011.
[442] Personal communication with Tyulkibas WUA's director, Vanovka, November 2012.
[443] Personal communication with the Tyulkibas WUA's director, Vanovka, November 2012.

making processes. From another perspective, the WUA's governing board was not able to involve the water users or to convince them of the relevancy of the WUA model. Therefore, it is possible to state that this new model, sponsored both by the international donors and the government, has not been adequately supported and understood by all the stakeholders at the district level. As pointed out at the end of Chapter 5, the government, instead of supporting the wide-spread adoption of the WUA experience through organizational and financial aid, has been widely supporting the Republican State Enterprises, in this case the one in South-Kazakhstan. The former director of the Tyulkibas WUA claimed that another possibility could be the RGP's control of water facilities, but this thesis was denied by the RGP members who stated that since the district does not include primary level canals or reservoirs, they cannot be directly in-volved.[444] Therefore the water resource management for the following years in Tyulkibas district could be a difficult enigma to contend with, since it is still un-clear and doubtful which organization can fill the institutional vacuum created by the crisis of the district water departments and the lack of understanding about the WUA model.

### 6.3.2 ORDABASY DISTRICT

Ordabasy district lies in the centre of South-Kazakhstan province in the middle Arys valley, north-west of the city of Shymkent. Surrounded by the steppes and by the Karatau range's southern slopes, the district's irrigated area reaches a to-tal of 32000 ha: most of it is irrigated by the Arys-Turkestan canal system. In the central section of the district, a few kilometres south of Temirlan, the dis-trict's chieftown, Arys canal arises from the river (on the right bank) and after 20 km flows into the Bugun reservoir. Close to the Arys canal's outlet lies another main facility, Karaspan canal, which irrigates 4200 ha. From the Bogun reser-voir (37 mil.M3 of water storage), arises the Arys-Turkenstan canal; built during the 1940s, it measures 97 km in length and irrigates a total of 55000 ha, provid-ing water to the district and also to the downstream district of Turkistan.[445] Since all the above mentioned water infrastructures are considered primary level, they are under the technical control of the South-Kazakhstan RGP *Iujvodkhoz*, spe-cifically of its branch based in Bogun village. In contrast, the secondary canals

---

[444] Personal communication with the Tyulkibas WUA's director, Vanovka, November 2012 and with the South-Kazakhstan RGP's members, Shymkent, November 2012.

[445] Personal communication with Ordabasy district water department' s members, Temirlan, April 2012.

which arise from the Arys and the ATK are the property of the Ordabasy district government (*Ordabasinsky Akimat*) and, therefore, parts of them are under the control of the Ordabasy district water department (*Kommunalni vodkhoz*) (FIG.30).[446]

## *A complex water context shaped by multiple approaches and rationales*

Whereas in Tyulkibas district the reform processes oriented towards the IWRM and related IMT implementation in the last years has been characterized by the shift from the district water department to the WUA and its consequent failure, in Ordabasy district the water context looks more complex and divided, including different actors managing water resources at the same time. Although different WUAs were formalized in 2004 according to the Water Code, the district water department is still working and involved in the water allocation procedure; in addition, the South-Kazakhstan RGP also plays a role at farm level. Hence, it is possible to affirm at a preliminary stage that the old post-Soviet management structure coexists with the new institutions promoted by the IWRM rationale. Whereas until 2004 the Ordabasy district water department managed the whole secondary canal network and ensured water allocation to the entire district's irrigated land (excluding lands lying close to the main canals), today, after the WUAs' formalization, the area under its control reaches 25.000 hectares, hence still the majority. However, the state organization, based in Temirlan, had several changes of staff and location in the last years: most of the former Soviet staff retired and nowadays, as emerged in Tyulkibas district, it is affected by a lack of technicians and experts. Furthermore, the district budget to support financially the authority is currently not sufficient to deal with canal maintenance and staff salaries. As it was pointed out in the analysis of the former Tyulkibas district water department, in Ordabasy too the farmers and the cooperatives sign a contract every year with the authority at the beginning of the cropping season. The important difference in comparison with the Tyulkibas unit is that the district authority deals only with the farmers which are not included in the WUAs. According to the debate highlighted at the end of chapter four regarding the district water departments at the national level, different and contradictory opinions have emerged about the current status of the Ordabasy water department: whereas both the director and the head of the district agricultural department (*Rayselkhoz*) stated that, despite lacks, they are carrying out their tasks and re-

---

[446] Personal communication with South-Kazakhstan RGP' Bogun branch, Bogun, November 2012.

sponsibilities, the South-Kazakhstan RGP staff claimed that due to financial issues and political decisions, the organization probably would be dismantled the next year.[447] The district authority, as emerged in the Tyulkibas district also, is not able to cover the costs for operation and maintenance of the district water department; therefore, according to the RGP Bogun branch members, the most likely strategy is that the secondary canals and the other district's infrastructures will be probably under the supervision of the RGP *Iujvodkhoz*, funded by the republican budget. The other possible path is that the WUAs newly established by the local water users will take control of those canals, but this option cannnot be easily realized: the governing board of Ordabasy district stated that for the farmers, the process of creating WUAs is very challenging, due to recent failures, and it strongly expressed its preference for state control, through the RGP, of water facilities and allocation.[448] Furthermore, the will to set up new WUAs should come from the farmers, and presently, as occurred in Tyulkibas as well, this process is hampered by a general lack of trust in the establishment of independent associations and by technical and financial issues.

---

[447] Personal communication with the Ordabasy district water department and agricultural department' members and with the South-Kazakhstan RGP' members, Temirlan and Shymkent, November 2012.

[448] Data collected through a meeting with members of the governing board of Ordabasy district, Termirlan, April 2012.

*Karaspan, Halik and Altursuu WUAs: a partial success of the initiative*

*FIG. 30: GIS elaboration of a satellite image (source: Google Earth TM), representing Ordabasy district (white shape) and the territories of the three WUAs established in the district (yellow shapes); in blue the river and the canals.*

Nevertheless, contrary to the other analyzed district, in Ordabasy unit since 2004 different WUAs were established by the water users, both according to hydrographic and administrative boundaries of the former *kolkhoz* and *sovkhoz*. Although some of them failed, three water associations are nowadays working, Karaspan, Halik and Altursuu WUAs, even though no development projects were set up in Ordabasy district by the international donors to support IMT processes, as for instance occurred in Machtaral unit lying in the southern part of

South-Kazakhstan province. In total, the irrigated area controlled by the three WUAs reaches approximately 8000 hectares, that is, on average, 25% of the Ordabasy district's irrigated area. Whereas from an institutional point of view those WUAs do not differ from the Tyulkibas WUA, from an organizational perspective they present significant differences, as they are totally independent from the district water department. The Karaspan WUA was established in 2005, but after years of organizational lacks, it was reformed in 2011 by the former Karaspan *sovkhoz* hydro technician and his son. Registered as a non-profit organization in the Justice Department, the WUA's irrigated area reaches 4265 ha, including two former *sovkhoz*; it is supplied through 18 secondary canals arising from the Karaspan canal. Those canals, own by the district government, are leased from the district water department for a term of 5 years. According to the WUA's staff and the farmers interviewed, water management and allocation significantly improved in the last year; farmers receive water according to the time schedules and thanks to the water fees (350 tenge / 1000 M3 -2.2 $) collected, new hydro-posts were recently installed between secondary and tertiary canals.[449] Besides these relevant technical improvements, focusing on the governance, the evidence has shown that the organizational structure is quite weak: the WUA does not have elected assemblies or farmers' representatives, but just organizes councils twice a year to discuss water distribution and agricultural issues. No elections have ever been organized for the WUA leadership. However, it is necessary to highlight that the Karaspan WUA, in contrast to other water associations, can rely on the knowledge and the experience, both technical and financial, of its staff: the accountant worked in the *sovkhoz* administration and the young director is the son of the first hydro technician of the former Karaspan *sovkhoz*, today employed in the Bogun branch of RGP *Iujvodkhoz*. The Altursuu WUA, established in 2004, shares similarities with the Karaspan WUA, even if it is not physically related to the Arys-Turkestanki canal system, but to the Guldriuk main canal, arising from Badam river. Here too secondary canals (12) are leased by the WUA from the Ordabasy district water department, irrigating a total of 1200 ha, the territory of the former *sovkhoz*. As emerged in the Karaspan WUA, both the governance, according to a participatory approach, and the organizational structure are weak and still far from the IWRM's participation principle nationally promoted by the international donors. Although the main staff organizes councils twice a year involving twelve farm-

---

[449] Personal communication with Karaspan WUA' staff and farmers, Karaspan, November 2012.

ers (the water users' representatives), the director, WUA's founder, has been the head of the association for eight years without any election. As he stated, the irrigation practices and the maintenance of the canals generally improved since 2004, so there will be no changes in the near future regarding the WUA staff. Furthermore, the Altursuu WUA director added that financial resources, fair management, and technical knowledge allow his WUA to work properly, but at the same time in the rest of the Ordabasy district other associations did not possess these requirements and failed.[450] The third association, Halik WUA — established in 2011 and linked to the Arys-Turkestan canal — is based on the hydrographic/administrative boundaries of the former *sovkhoz*. According to the director of the Jenis village land office, Halik WUA was established to deal with the mismanagement and lacks of the Ordabasy district water department regarding secondary canal maintenance and water allocation. In the last two years, with the financial resources gathered through the water fee collection, they were able to start the restructuring of secondary canals and hydro posts.[451] The interviewed farmers claimed that for some years it has been very difficult to deal with the district water department because of their lacks and neglect in water allocation. Moreover, they added that the WUA's formalization was possible thanks to the organizational and technical skills of the heads who worked in the *sovkhoz*.[452]

### A different reality: the relations between the water users and the South-Kazakhstan Republican State Enterprise

After an overview of the WUAs currently working in Ordabasy district, it is possible to state that their success is strictly connected with the technical and organizational skills of their heads and staff, more relevant than the attitude and behaviour of the water users. In this context, the WUAs heads — despite deficiencies in the participation of water users and their lack of involvement in the decision-making processes — were able to convey to the users the importance of water payment (ISF) in order to ensure the subsistence and equitable performance of the WUA. In contrast, as discussed at the beginning of this section, the IMT process has not involved the farmers who own the land close to Arys and Arys-Turkestan canals. Since no secondary canals have to be managed and

---

[450] Personal communication with Altursuu WUA' director, Shymkent, November 2012.
[451] Personal communication with Jenis village' Land office, Jenis, November 2012.
[452] Personal communication with the Halik WUA's farmers, Jenis, November 2012.

maintained in that area, those farmers request water directly from the South-Ka-zakhstan RGP's Bogun branch. In addition, they have never been connected to the Ordabasy district water department, so due to the fair allocation provided by Bogun RGP branch, in the last ten years no WUAs have been established.[453] According to this authority, those farmers have more certainty regarding water allocation procedures, being involved with a state enterprise; furthermore, he added that the farmers avoided all the mismanagement and lacks related to the district water department's financial and operational shortages as well as the problematic and challenging WUA establishment process which often requires organizational skills and financial resources that are hard to get.

### 6.3.3 OTRAR DISTRICT

Otrar district lies inthe downstream part of the Arys valley, on the western side of the South-Kazakhstan province. Most of its territory includes steppes and deserts and only a small area is suitable for irrigated agriculture, due to the canal systems; the district is crossed by the Syr-Darja, the Arys, and the Bogun rivers. From the 1940s to the 1970s, a few kilometres upstream where the the Arys river flows into the Syrdarja, a canal system has been built arising from the Shaulder dam, including Altymbekov Shaulder and Kokmardan canals. This water system increased the total irrigated area up to 16.000 ha. In its northern part lies another irrigated area, Aktyube-Celik, ranging 4.500 ha; in contrast to the Shaulder irrigated area, directly supplied by the Arys river, this one is irrigated by the Bogun river, a natural watershed regulated upstream by the Bogun dam, in Ordabasy district (FIG.31). However, in the last years, due to soil salinization problems and water shortages, only 75% of the total irrigated lands were available for agriculture. As already analyzed in the other districts, the main canals and infrastructures are under the control of the South-Kazakhstan RGP *Iujvodkhoz*.

### The WUA's crisis and the uncertain water management context

The evidence has shown that in Otrar district the reform processes oriented towards the IWRM are proceeding differently as compared to both Tyulkibas and Ordabasy; this context shows how nowadays these processes are complex and

---

[453] Personal communication with South-Kazakhstan RGP Bogun branch' members, Bogun, November 2012.

varied depending on the district's characteristics and related local realities. Nowadays only one WUA has been working while the rest of the secondary level water facilities are under the Otrar district water department's control. According to the director, although until the 2011 water shortages' issues occurred, being at the tail end of the river and due to the facilities' conditions, during the winter 2011/2012, the Altymbek canal and part of the secondary canals were restored within the province's budget and new hydro posts were installed improving water measuring and supply.[454] In 2012, 1200 water users (independent farmers and cooperatives) signed a contract with the district water department, paying an average of 195 tenge /1000 M3 (1,3 $); the fee is lower in comparison with Ordabasy district, since the water is deviated directly from the river and subsequently there is no additional fee for the Bogun/Arys-Turkestan canal system.Although the staff stated that the water department works equitably and is able to provide equal water allocation to the farmers, this statement was contested by several water users.

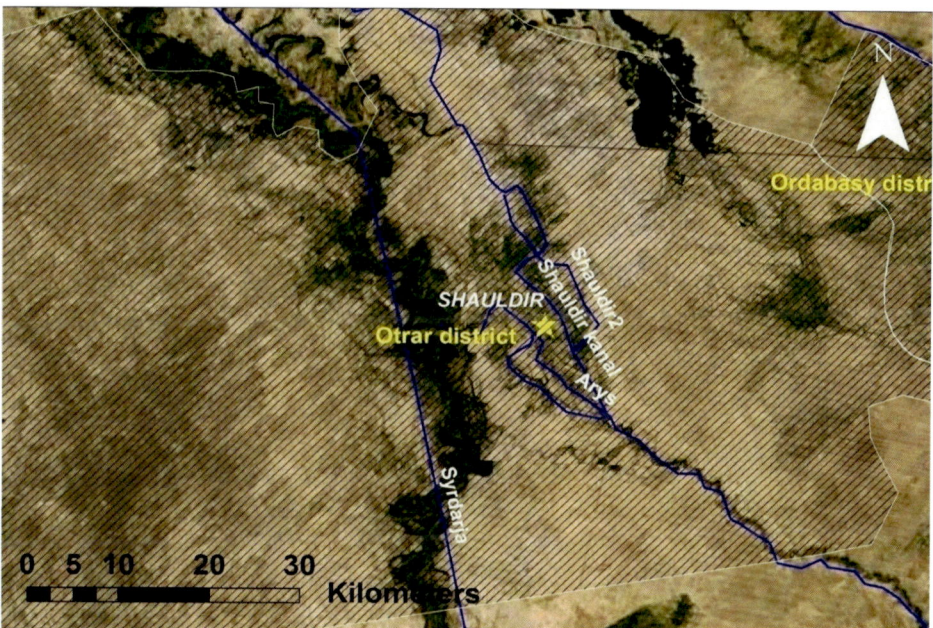

*FIG. 31: GIS elaboration of a satellite image (source: Google Earth TM) representing the Otrar district's territory (included in the white shape), Shauldir town (yellow star) and its canals' system (blue lines).*

---

[454] Personal communication with Otrar district water department' director, Shaulder, November 2012.

According to the small-medium plots owners (7 to 15 ha), the water department does not respect time schedules, part of the *miraab* do not work, and water supply inequities are widespread among small and large farmers; furthermore, some of the farmers receive benefits, as they are in close or familiar relations with the district water bureaucracies.[455] Since 2003, when the WUAs' law was issued, the reforms' process began, but after a few years it was slowed by management and organizational issues; two WUAs were established (in 2008), but after two years they were dismantled due to lack of financial, organizational, and technical skills. According to the district water department's director, the WUAs did not have enough funds to support themselves, and to carry out the operation and maintenance of the canals. Moreover, some farmers interviewed about the WUA experience confirmed that the water fees had increased since the WUAs were working, without significant improvements in water allocation procedures. In addition, regarding the organizational structure, important changes did not occur; hence most of the farmers stated that they would prefer to deal with the district water authority, despite some lacks.[456] As emerged in the other two districts, in Otrar district the water context appears unclear and unstable for the near future; the reforms processes oriented towards the IMT are not completed and not properly supported and the district departments are, as already pointed out, in a precarious position. Although no information about the current district financial budget for water management was locally released, the South-Kazakhstan RGP *Iujvodkhoz* members affirmed that the district authority (*Rayon akimyat*) is no longer able to fund the district water department's operations.[457] Therefore, the most feasible option is for the secondary canals to be placed under the responsibility of the RGP *Iujvodkhoz*. According to their members, the state budget for water infrastructures nationally increased in the last two years, hence the shift in management could be easily conducted. Nevertheless, the national Committee of Water Resources and also the Aral-Syrdarja River Basin Agency have not participated in the discussion regarding the next water management context in Otrar district and they did not provide any information about which authorities would be involved. As pointed out in Ordabasy district, the hypothetical establishment of new WUAs has been recently considered a difficult and challenging process, due to the issues which have affected some of the associations — such as not

---

[455] Personal communication with the Otrar district farmers, Otrar district, April and November 2012.

[456] Personal communication with Otrar district farmers, Otrar district, April and November 2012.

[457] Personal communication with South-Kazakhstan RGP's members, Shymkent, April 2012.

enough support from both the district water department and most of the farmers interviewed.

### *The unceratin future of the WUAs; the potential failure of the IWRM oriented reforms*

As mentioned above, only one WUA, named Mahambet/Aktyube, is currently working, since 2007, in the northern part of the Otrar district, supplied by the Bogun river (FIG.32). Due to this main water course, not directly related with Arys river and the Shaulder canal system, the Mahambet WUA has been operating since the end of the 1990s, when it was founded as an informal water users association (*Associazija Vodopolzovatelii* or AV), in connection with the RGP Bogun branch instead of the district water department. Based on the Aktyube *sovkhoz* administrative unit, comprised of 2500 ha, the WUA can rely on the competencies of its staff, which worked on the governing board of the Aktyube *sovkhoz* during the Soviet Union. Officially the Mahambet WUA started working and was registered in the justice department, according to the Water Code, in 2007 when it shifted from a AV to a water users association. The farmers interviewed stated that in this territory water division and allocation are carried out in a better way when compared to the Shaulder area; the number of water users is lower and though no councils have been set up in the last years—neglecting, therefore, to adopt a participatory approach — the WUA generally complies with its responsibilities. The water users added that since the irrigated land is not large and lies in one village, they informally meet each other to discuss potential issues and therefore no kind of WUA councils have ever been created.[458] According to the Aktyube village land office, the Otrar district water department recently prepared documents asking for the dismantling of Mahambet WUA and the end of the secondary canals' leasing contract — despite the WUA's current fair practices previously analysed. This can be seen as a senseless move, considering the current conditions of the district water authority.[459] This request, clearly considered and understood as a rebuke by the district department against the WUAs, was refused by the director and, as emerged, the disputes would be taken up by the Otrar district's court and by the South-Kazakhstan RGP *Iujvodkhoz*.

---

[458]Personal communication with Mahambet WUA' water users, Aktyube, November 2012.
[459] Personal communication with Aktyube Land Office' members, Aktyube, November 2012.

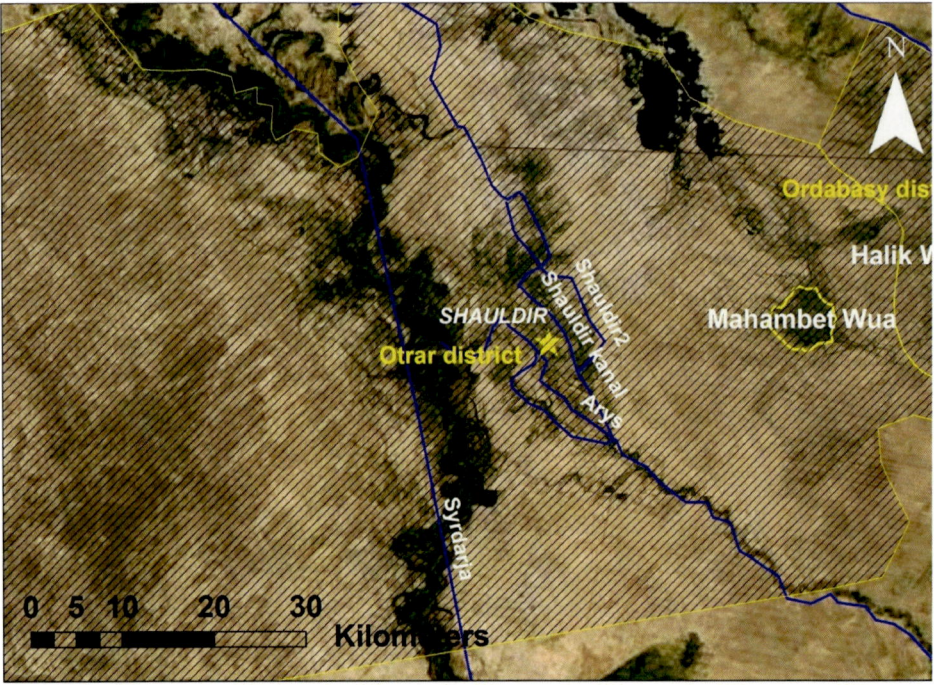

*FIG. 32: GIS elaboration of a satellite image (source: Google Earth TM) representing the territory of Mahambet WUA (small yellow shape) (eastern part of Otrar district –large yellow shape); this irrigated area is not supplied by the Shaulder canals' system. Blue lines: rivers and canals.*

Therefore, reflecting on the Otrar district context, it is possible to state that this unit too is affected by institutional and organizational instability and that the forthcoming water management processes are still uncertain. Given the financial issues of the district water department and the lack of trust and support for the WUA model, the South-Kazakhstan RGP seems in a favourable position regarding the forthcoming water management context. Therefore, the evidence clearly shows a process of distancing from the idea of developing and strengthening the IMT and the IWRM framework, in contrast with the ongoing international donors projects and with the national measures enacted only ten years ago.

# 7. THE LOGICS OF THE IWRM IMPLEMENTATION IN CENTRAL ASIA: COMPARING THE UZBEK AND KAZAKH EVIDENCE

## 7.1 INTRODUCING THE COMPARISON

The previous two chapters offered a detailed description and analysis of water reforms implementation and related issues at the basin and local level, according to the IWRM rationale, in two case studies, the Middle Zeravshan valley (Uzbekistan) and the Arys valley (Kazakhstan). This chapter focuses on the comparison between them, highlighting the similarities and differences, in order to get a clear, in-depth understanding of the issues and challenges related to the IWRM implementation path and the related IMT success in the Central Asian region. According to the analysis, data will be compared across multiple scales: the first section compares the water management authorities at the river basin level, highlighting the similarities and differences in relation to the IWRM pillars widely discussed in Chapter 1; in addition the emerging "conflicting points" will be discussed. The second section sums up the reform experience in the two case studies, focusing on the water authorities and related issues at the local level, WUAs, and district water departments. In order to have a clear overview and understanding of the IWRM pillars implementation, the following will be singularly discussed: institutional and organizational structures; changes in boundaries towards hydrographic principles; integration and participation of the stakeholders; and the introduction of monetary economic systems.

## 7.2 COMPARING THE BASIN LEVEL WATER REFORMS

According to the IWRM rationale, widely discussed in the first chapter, water management should be based on hydrographic principles in order to achieve the greatest efficiency and respect the environmental units. In Uzbekistan at the basin level, the shift from water management based on administrative principles to the new hydrographic units occurred in 2003, as discussed in Chapter 4, through the enactment of Decree n.320: this measure replaced the 13 former province water departments (*Oblastvodkhoz*) with 10 Basin Irrigation Systems Authorities (BISAs). However, when the territorial characteristics of the new entities are analysed in depth, it emerges that only five of them are effectively based on hydrographic principles; the other five have kept the former administrative boundaries. Reflecting on this institutional change, it emerges that the Ministry of Agriculture and Water Resources was able to profit from the donor-sponsored reform according to the IWRM pillars, changing the institutional water structure for internal political reasons. Although the official purpose was to create hydrographic units, the Ministry, in an effort to abide by the measure of dismantling the province water departments, wanted to reduce the province government's power in influencing water control, according to a recentralization rationale. Yalcin and Mollinga (2007) also state that the real aim of this national reform was to reduce the power of the province government leading the BISAs' control under the Ministry in Tashkent; therefore, according to Yalcin and Mollinga, the former Minister Djalalov was able to enact a reform oriented towards an institutional and organizational recentralization, though based on basin principles, with the support of the donors and of the ICWC.[460] Hence it is possible to state that the hydrographization of basin level units was only halfway achieved. Furthermore, from the data analysis, it emerged that no changes regarding the staff and the organizational tasks occurred and, therefore, for half of the newly established units the enactment of this decree was only a change in terminology (name of the institution) . Also the non-compliance to create a structure oriented towards a participatory approach in the decision-making processes — in this case, the basin councils — which is strongly promoted by the donors and at the base of the IWRM rationale, clearly demonstrates a resistance to changing the governance and organizational environment of the BISAs. This conservative approach emerged both from the heads of the governing boards and the other members of the staff which did not support a participatory approach in the decision-making processes. The attitude of the BISAs' members allows to understand the significant hesitance to change the existent social structures which inevitably affect the institutional ones and directly characterized their perspective on the IWRM rationale. The Zeravshan Irrigation Basin Agencies, though one of the five based on hydrographic boundaries since 2003 (the former Samarkand province water department plus other neighbouring districts) has not effectively changed its organizational structure, nor established the basin councils or other forms of participatory approach in decision-making procedures. Differently, in Kazakhstan, which is characterized by a different water institutional framework

---

[460] YALCIN R., MOLLINGA, P., 2007. "cit.".

since the Soviet Union (Committee of Water Resources instead of the Ministry), the authorities based on basin principles (river basin agencies, BWO) were inherited and therefore formalized through the 1993 Water Code as the main agencies involved with water management. Nevertheless, as it was discussed in full in Chapter 6, their tasks were shared with the province water departments (the Republican State Enterprises since the end of the 1990s) and often conflictual. According to the enactment of the new water code in 2003, based on the IWRM framework, the river basin agencies were institutionally reinforced and oriented more towards the control of the water use, (quality and quantity). Specifically, the strengthening of these authorities was oriented towards the integration of water use (irrigation, domestic, energy, and industries) and the promotion of a participatory approach in water management. In fact, due to a donors-based project (UNDP and the government of Norway), since 2007, the basin councils including multiple stakeholders were established in all the basin agencies. Though, on the one hand, the concept of integration is developing quite well, on the other hand, the organization of the basin councils and the related participation of the water users — after the initial widespread acceptance of these principles — is still weak and fragmented, particularly after the first stage of the donors' project ended. According to the data, the meetings were not regular, some potential members decided not to participate, others were not involved, and others gained a prominent role due to political reasons. Notwithstanding, as it was widely debated in relation to the IWRM, in recent years (since 2010) the Republican State Enterprises, having a more technical role in water allocation, have been more supported, financially and politically, in comparison with the River Basin Agencies, particularly regarding the local issues and relations with the water users. In addition, the Republican State Enterprises, having kept the same structure as the province water departments, were not induced to change their organizational structure nor particularly inclined to introduce a participatory approach in their decision-making processes. This current scenario appears controversial and in conflict with the IWRM framework implementation process and the new water code, as it gives more governmental support to the entities based on the former Soviet institutional and organizational structures. Although examples of participation were initiated and some basin councils established, the evidence clearly shows that water management practices at the basin level are still characterized by a top-down approach and a strong hierarchic structure. These attitudes in water management processes do not significantly differ from Uzbekistan, even if in the neighbouring republic they are even stronger and in addition, it should be underlined that no kind of participatory approach in decision-making processes has ever been supported by the government. Focusing on the relations with the donors, regarding the basin level, the results have shown that both countries accepted to collaborate with them and to undertake a reform process of their water sector. As it was discussed in the first chapter, on the basis of Molle's (2008) reflection, this political strategy at the national level — that is, the acceptance of the donors' and the development agencies' initiatives — allowed them to receive several benefits, such as considerable amounts of money and loans, and to organize meetings and conferences with international experts.[461] Furthermore, Uzbekistan and Kazakhstan's collaboration with international donors allowed the legitimization for the reconfiguration of their water bureaucracies and water policies, and enhanced their reputation in the international water community. However, comparing the actions and policies undertaken by the two countries, quite significant differences emerged. The Uzbek government has not enacted a new water code, avoiding the IWRM formalization; also, it has limited the reform path to merely launching the decree which led to the reconfiguration of the water authorities at the basin level. However, this institutional change only focuses on the hydrographization of the water authorities — though partial and incomplete, as discussed above —without considering the other pillars of the IWRM. In contrast, the Kazakh government — notwithstanding the issues previously mentioned and an ambiguous action towards the sponsored reformsin the last years — formalized the IWRM framework (the first republic in the Central Asian region to do so) and used this as the basis to develop the concept of the integration of water use and a participatory approach in decision-making processes. Through the issue of the Water Code formalizing the IWRM, Kazakhstan aimed to gain a leading role in water reforms in the regional scenario and consequently the reputation of the most open republic in regards to political and economical changes as well as in relations with development agencies and donors. Even though the government gained this role and related benefits, as emerged in the last years, the reforms process has not been completed yet. It seems quite clear that the government-sponsored reforms were mainly used as a political strategy to attain a dominant role.

---

[461] MOLLE, F., 2008. "cit.".

# 7.3 COMPARING THE IMT AND THE IWRM IMPLEMENTATION AT THE LOCAL LEVEL: THE WUAs INITIATIVE

As post-Soviet republics, the challenges for water management at the local level are similar in Uzbekistan and Kazakhstan: both countries had to cope with deteriorated infrastructures, decreased financial resources and technical knowledge, and hierarchical governance systems that are often not adequate to face the new challenges as the arise of the peasant farmers. To face with these issues both countries had to develop new strategies and policies, under the donors pressure to implement the international norms as the IWRM the related IMT. According to the empirical findings from the two case studies, discussed in-depth in the previous chapters, significant differences in the institutional/organizational structures have emerged. From a comparative perspective, as a primary point, the analysis has shown that in Middle Zeravshan valley, Uzbekistan, only one typology of authority is involved with water management — that is, the donors-supported Water Users Associations; while in Arys valley, Kazakhstan, both the WUAs and the district water departments inherited from the Soviet Union are both actively involved. This initial and significant difference brings up several questions that will be answered in relation to the analysis of the IWRM pillars.

## 7.3.1 The WUAs: new forms of state control Vs. independent -weak- water organizations

As it was analyzed in Chapter 5, according to an institutional perspective, in Uzbekistan the shift from the district water departments and the *shirkat* to the WUAs was first discussed at the end of the 1990s and therefore was included as the second step of the "Program of measures on the improvement of irrigated lands for 2001–2010". Subsequently the Decree n.8 of 2002, in absence of a law, was the first legal measure promoting the WUAs establishment process which has been widely adopted since 2003 due to the support of the international donors, specifically the World Bank and the Asian Development Bank. After some years, in 2009, according to the IWRM pillars supporting hydrographization and participation in decision-making procedures, the government enacted a law (introducing amendments to the 1993 law on water and water use) promoting hydrographic-based WUAs directed by water users instead of by members of the state organizations. Nevertheless, although the WUAs became widespread throughout Uzbekistan beginning in 2003, a law formalizing them has never been enacted; nationally, they should be registered as associations of water users in the province justice department. However, as fully discussed in the previous chapters, the WUAs were not funded by free initiatives from the farmers but formed as a result of the decisions of the former directors of the district water departments or by members of the basin level state organizations. This was a clear move to maintain the system of state control of water allocation and agricultural production, according to the state quota system for the most important crops (cotton and wheat). On the other hand, in Kazakhstan the move towards the establishment of the WUAs started in the second half of the 1990s due to the support, in some areas, of the international donors (World Bank, Asian Development Bank, and USAID). Contrary to Uzbekistan, the Kazakh government — in the wave of water reforms supporting the IWRM framework — enacted Law 404-II, formalizing the WUAs as Rural Consumers Cooperatives of Water Users (*SPKV)*, thereby bestowing an official status to the already existing associations and giving the farmers a chance to self-organize a WUA. Therefore, a significant difference emerged between Uzbekistan, where the WUAs establishment were co-opted by the state organizations, and Kazakhstan, where the farmers — even though, in most cases, they had been the former heads of the collective farms — were not influenced by the state authorities in the establishment of the WUAs. As stated at the beginning even though, on the one hand, in Uzbekistan only the WUAs are involved in water management and allocation, somewhat under the control of the state organizations; on the other in Kazakhstan, besides the WUAs, the district water departments, although with the discussed financial shortages and organizational lacks, are still operating. Therefore the governmental strategies of water control at the local level in the two case studies present differences. As discussed in Chapter 5, in the Middle Zeravshan valley two models of WUAs emerged: one based on local initiative, which largely prevails, and the donors-based model, exclusively in the Pastdargom district. The state authorization to set up the donors project and the new WUAs in one district allowed the government to maintain a local level water structure which was not significantly different from the Soviet experience — since the creation of the WUAs and the current performance is still state controlled — and, simultaneously increase its international reputation in relation to the international agencies. In contrast, in the Arys valley the government allowed the free establishment of the WUAs, implementing the issued law and in turn increasing also its reputation, and from the other kept operating the Soviet inherited water structure to provide water allocations to the farmers who do not have the possibility to create an association. Nevertheless, afterwards, the partial crisis of the WUAs experience and the financial crisis within the district water departments initiated the re-centralization process supported through the action of the Republican State Enterprise. Therefore it is possible to state that, according to an institutional perspective, the introduction of the WUAs in Uzbekistan, at least in the case study, led to only a nominal change in the former management structure, keeping the original water political hierarchy and related procedures in place. In the Kazakh case study, however, the WUAs experience initially led to changes oriented towards decentralization and a fragmentation of the local-level water structure, decreasing the state control; afterwards the partial failure of the WUA legitimized the state's reconsideration of centralizing processes.

### 7.3.2 Reorganizing the water authorities: local reinterpretations of the hydrographic principles

During the Soviet Union, water resource management at the local level, as well as the province and national one, was based on the administrative boundaries of the district water departments and the collective farms. As it was fully discussed in Chapter 1, the IWRM framework promotes hydrographic - based administrations as the best institutional structure, which also take into consideration natural and environmental concerns. In Uzbekistan, since a new water code based on the IWRM principles was not enacted, the hydrographic management principle concerning the WUAs was introduced with the previously mentioned law of 2009. In Kazakhstan, since the IWRM was formalized, the law on WUAs clearly mentions that the newly established associations, as well as the former ones, should be based on the hydrographic principles. Therefore, both countries, although with different legislative approaches, officially supported the new management principles concerning the WUAs. Nevertheless, especially regarding the already established associations, a management principle shift requires significant changes in the hierarchical structures and related local political powers. In fact, an organization based on hydrographic principles would be less affiliated to local authorities and hierarchies and politically more independent. Probably for these reasons, although both countries officially supported the shift, the implementation process shows similarly lacks in both case studies. In the Uzbek case, excluding the Pastdargom project based area, the WUAs still refer to the boundaries of the district water departments and hence no changes have occurred since the collapse of the Soviet Union. Besides resistance to changes in the local political nomenklatura, who still sit at the head of the WUAs, the territories of the associations are further subdivided in the former collective farms (*agrofirma*) of which the former managers still provide to the farmers technical aid for cotton cropping. In Pastdargom district, the subdivision of the former WUA into six new ones based on hydrographic or coinciding administrative - hydrographic boundaries led to some changes in organizational structure and brought millions of dollars into the province government's budget. In the case of Kazakh, although the above-mentioned law on WUAs, the associations — having been created inside the territory of the district water departments — refer to the administrative or the coinciding administrative-hydrographic boundaries of the former collective farms, and in Tyulkibas to the district boundaries. Hence no real efforts were made by the local governors to genuinely support the shift to hydrographic boundaries. Therefore, in both case studies, the implementation of the shift was similarly partial and weak and hampered by an unwillingness to challenge and change solid political structures.

### 7.3.3 Supporting centralization!: the failed shift from a top-down to a participatory approach

As it was fully discussed in Chapter 1, the IWRM framework was created — besides being impelled by the desire to improve water resource management according to a sustainable approach — to support decentralization and the establishment of democratic principles within the sociopolitical system. Both Uzbekistan and Kazakhstan inherited the socio-political structure of the Soviet Union, characterized by a hierarchical model with strong centralization and a top-down approach and a lack of horizontal coordination; this model still someway shapes the current socio-political culture, although with differences between the two countries. In the Uzbek case studies the evidence has shown that a real participatory approach in the WUAs' decision making is still far from widespread. As it was fully analyzed in Chapter 5, no sort of WUAs council or horizontal governance was established. On the one hand, part of the farmers still perceive the WUA as an organization shaped by the state hierarchical structures and hence based on the Soviet inherited model, while others clearly expressed that participation in decision-making processes has not led to any advantages in water allocation or in the fulfilment of the state crops plan. Moreover, other water users did not have the perception of any institutional changes — for example, they still call the WUA *rayvodkhoz* (district water department) and, preferring not to be in close relations with the local political powers, they avoid any sort of participation in the decision-making processes. On the other hand, the heads and the members of the WUAs, despite the action of the donors to influence the local procedures, are still supporting a top-down approach due to the characteristics of the Uzbek sociopolitical model, which is not so distant from the inherited Soviet one. For these reasons the local water nomenklatura has not tried to support a change towards a participatory approach nor widely shared the lessons learned from such an approach to the water users. In this connection, the water users themselves, due to the national sociopolitical environment, have never expressed the willingness to be involved in the decision-making procedures, trying to avoid potential disputes both with the state authorities and other water users. Even in the Pastdargom donors-based WUAs, where kinds of councils are somewhat experiencing, it is not considered by the local hierarchies a potential model to be widely shared among the other WUAs. In contrast, in Kazakhstan, the government, induced by the previously described donors projects, supported the establishment of the basin councils and sub-basin councils where the members of the WUAs and the district water departments could participate. Nevertheless this initiative has not really changed the governance: some farmers claimed that these councils are an opportunity for the local water hierarchies to meet, without any real involvement of the water users; others stated that no elections for the governing board had ever been organized and the process is still characterized by a top-down approach. Furthermore, other water users were not really interested in the initiative, nor in actively participating, in general, in changing the governance procedures. In the WUAs of Ordabasy and Otrar districts, the staffs organized some councils, but in these cases as well two issues emerged: firstly, the conservative approach of the governing boards led to no elections and few possibilities from the water users to really influence the decision-making processes; secondly, the widespread hesitation to participate — in particular, by the smaller farmers, due to the afore mentioned conditions and related procedures — was somewhat influenced by the farm-level political issues. Therefore, whereas the Uzbek local hierarchies have shown the willingness to keep a top-down approach without questioning the former governance system, in contrast the Kazakh hierarchies, although with the donors' pressure, gave the water users the possibility to change the top-down governance structure by introducing democratic mechanisms. Nevertheless, in both the case studies, the evidence has similarly shown how, from the perspective of the water users, the Soviet-inherited way of thinking — in terms of receiving a service in a vertical manner without questioning and/or limiting the interference of the state authorities and local hierarchies in decision-making processes — is still widespread.

### 7.3.4 The -failed- commodification of water

The IWRM and donors-supported establishment of water fees (ISF) presented a significant change in Central Asia, where water has never been paid nor ever been considered an economic good. Instead, historically and traditionally water has always been considered a gift of God; during the Soviet Union it was allocated by the state agencies in a top-down manner on the basis of agricultural production plans. Therefore from the perspective of the water users, the payment of water fees represented a big challenge. The expectation was that the ISF would lead to efficient water use, cost recovery for the WUAs, and the reduction of water waste. Nevertheless, in Uzbekistan the ISF, though informally introduced in 2004, is in conflict with the state production plans. Since the cotton and wheat production plan must be fulfilled, the WUAs still provide water to the farmers even without having collected the fees. The farmers, on the one hand, have expressed resistance to paying the fees since water has never been considered a market commodity and their own financial shortages make it difficult; on the other hand, since the WUAs are not financially independent but subsidized by the state organizations, they are somehow able to operate without a real cost recovery. It is important to consider that due to the deteriorated condition of the water infrastructures it is also challenging to count on the real water use. Therefore, the ISF is paid approximately per hectare, the WUAs do not recover their allocation service expenditures, since according to the evidence of this case study, less than approximately 20–30% of the collection rate can be counted on. The context slightly differs in the Pastdargom district, where donors' subsidies contributed to the installation of measurement points to calculate the water flows and through their action, the ISF was more effectively supported in the new WUAs. Nevertheless, their staff stated that the water fee collection rate is still low and the widespread collection of the water payment fee remains a significant challenge. In contrast, in Kazakhstan the water fees had been already introduced in 1997, a procedure supported by the donors' actions, during the first WUAs' establishment process. The strengthening of the ISF occurred in 2009 when the Committee of Water Resources fixed a national water price according to cubic metres, to which the different WUAs add an additional fee for cost recovery depending on the water networks and the often obsolete infrastructure. Nevertheless, contrary to the government's mandate to support the ISF, in the

Kazakh case study most of the water users resist the water fees collection, similar to the Uzbek situation, citing various reasons — such as the departments are not performing a service and financial shortages; concerning the WUAs, an increased fee in comparison with the one paid to the district water departments. These dynamics, widely discusses in the previous chapter, led to the inability of WUAs to recover their costs and the related failure of several water associations in the last years. The case study evidence shows only 50–60% of the full water fees were collected. These issues also explained the preference, for a considerable part of the water users, to pay less and receive the services from the state organizations instead of supporting a change which will include an active participation in local water management. Therefore, though the government strategy for the ISF considerably differs between the two case studies, the resistance of the water users to pay the fees, particularly the increased ones, is influenced more by the total absence of economic principles for several decades than, where established, by the donors-based projects.. Having analysed and compared the IWRM pillars in relation through the two case studies, the following table summarizes their implementation:

| IWRM / IMT | UZBEKISTAN | | KAZAKHSTAN | |
|---|---|---|---|---|
| | **Basin level** | **Local level** | **Basin level** | **Local level** |
| Integration | **no** | **no** | **yes** | **no** |
| Hydrographization | **partial (50%)** | **no** | **partial** | **partial** |
| Participation | **no** | **no** | **Yes (weak)** | **partial** |
| Water fees | **Yes/ "on paper" and very weak** | | **Yes / weak and partial** | |
| WUAs (according to the IMT) | **Yes (Uzbek typology)** | | **Yes ( weak and part of them failed)** | |
| Bottom-up practices | **no** | **no** | **partial** | **weak** |

# 8. THE TRAJECTORIES OF THE IWRM IN CENTRAL ASIA: REDISCUSSING THE FRAMEWORK

## DISCUSSION AND CONCLUSIONS

The last chapter discussed and analyzed the comparison of the empirical findings of the two case studies in relation with the IWRM pillars and the IMT process across multiple scales, focusing on the basin and the local levels. The table above summarizes the results of the IWRM implementation process in Uzbekistan and Kazakhstan, allowing a reflection on and understanding of these dynamics in the Central Asian region. Coming back to the initial research questions, after the presented analysis and comparison, it is possible to answer them: Which trajectories regarding water policies have been undertaken in Uzbekistan and Kazakhstan in the last decade? What are their current paths towards the IWRM implementation? Finally, in terms of the current global water paradigm, has the IWRM framework, as promoted by the donors, been implemented in Central Asia, or has the sociopolitical environment in this region shaped and influenced the process and its dynamics? What logics have emerged? The first significant difference which emerges between the two countries concerns the national institutional water framework: in Uzbekistan, although in the last years, the amendments to the 1993 law on water use that were issued somewhat support the IWRM rationale, a new water code or law formalizing the IWRM framework has not been enacted yet. In Kazakhstan, the IWRM was formalized through the 2003 new water code. Nevertheless, as discussed, this institutional difference is less significant concerning the government procedures that have occurred in the last years. Despite the significant institutional difference, a strong similarity emerged among the two case studies: both the governments, in the IWRM implementation process, focused on and chose to support the most convenient pillars, implementing them according to a national interpretation which did not interfere with their political and economic systems. Both countries, although with differences, preferred to maintain a conservative approach, in terms of limiting the institutional changes and in implementing reforms without questioning their governmental systems. It is clear that a full IWRM implementation would have required major changes in their respective governmental structures, district and local hierarchies, sociopolitical procedures and relations within the civil society. Focusing on the WUAs establishment, the results significantly differ between the Middle Zeravshan valley and the Arys valley. In the first case study, except for the donors-based project area, the analysed WUAs represent a local reinterpretation of the former Soviet local water framework supported by the local government hierarchies, involving mostly just a change in names. In Kazakhstan, the WUAs have been established freely by the water users without any interference of the government according to the IMT rationale. Nonetheless, similarly a top-down approach between the government and the water users, and between the WUAs' governing board and the water users, in the decision-making processes is still present — although in Uzbekistan it is even stronger. Therefore, bottom-up practices have not emerged. Subsequently, a participatory approach has similarly not been widespread, although the Kazakh government, in contrast to the Uzbek government, tried to support it among the water users and in the relations between the water users and the governing boards of the WUAs. Concerning this issue, the behaviour of the water users, in some cases their clear resistance to participating in the decision-making processes, is partly inherited by the procedures which characterized the Soviet sociopolitical system and partly due to local political issues between them and the state organizations. Therefore, whereas Uzbekistan in the last decade has a kept a strong state-centralized approach in its sociopolitical structure and in water management procedures, Kazakhstan, after a decade of supported reforms and a slight shift towards decentralization (from the late 1990s through 2010), in the last years has undertaken a nationally based re-centralization process, as analyzed in depth. Therefore, in response to the research question "What are the logics which affect the IWRM framework implementation in Central Asia?" it is possible to answer that the current IWRM implementation experience in the Central Asian region is a very challenging process, and according to the results, it has been strongly affected by the national sociopolitical rationales. Therefore the framework, as promoted by the donors, is somewhat in crisis and is far from being completely implemented.

## 8.1 THE UZBEK WAY TO THE IWRM CONFLICTS WITH THE IWRM RATIONALE

The IWRM rationale and the connected IMT process is totally in conflict with the Uzbek system, its sociopolitical structure, its economic system, and with the state governmental approach in water resources management. The Uzbek government clearly expressed resistance to questioning and trying to change its sociopolitical and economic system, a required condition for achieving a full IWRM implementation, with the exception of the donors-based project areas (which support the IWRM) — in the Fergana valley at the national level and the Pastdargom district at the local one — where the government used their approval of the IWRM framework as a strategy for enhancing their international reputation, while keeping and preserving the former system. In fact, the case-study results have shown that the government and the local state organizations have tried to implement only the parts of the pillars which would not require major changes in the local level sociopolitical structures and procedures. Furthermore the government, through the 2003 decree on "Restructuring of national water management", was able to achieve a recentralization process — taking water control from the province departments and giving it directly to the ministry — by establishing the new irrigation basin agencies (BISA) based on hydrographic boundaries (but, in the end, only half of them). In addition, by showing the establishment of the WUAs to the international donors, and collaborating with them in the project areas, the government was able to keep the state quota system inherited from the Soviet Union operative (100% of the state quota for cotton, and 50% of the wheat quota), limiting the farmers' freedom and the water users' participation, which is totally in conflict with the IWRM framework rationale supporting water conservation and asserting the economic value of water. Persisting with the state-quota system is the best strategy for keeping control of water and agricultural production under the state-centric model according to a vertical top-down approach; that is, controlling water and agriculture as a means for limiting the autonomy of the water users and the farmers. Therefore, on the one hand, the government was able to profit from the IWRM to legitimize its aim and strategies — keeping a state-centric system and maintaining sociopolitical control of water resources management in a way that is not so different from the Soviet experience — while on the other hand, the water users and the other stakeholders at the local level did not have the possibility , nor the will probably, to demand a change towards a participatory approach in the decision-making processes. Hence the Uzbek government, by partially supporting the IWRM framework, was able to receive a great deal of money through loans and project establishment as well as gaining a reputation for being a "collaborative country" with international agencies, without any intention of  introducing any sort of democratic principles into its stable and conservative sociopolitical system. An effective and whole implementation of the IWRM framework in Uzbekistan would require a radical change of the entire sociopolitical system inherited from the Soviet Union and the related hierarchies, the cessation of the state quota system and an authentic shift to a market economy. These processes should also be accompanied by a radical change of the involvement of the civil society in the local decision-making processes and in everyday life in relation to the local powers — changes that are not expected soon. At the same time, it seems that currently the Uzbek government does not have the willingness or plan to deeply question its system and undertake a change.

## 8.2 THE KAZAKH WAY TO THE IWRM: A NATIONAL REINTERPRETATION OF THE FRAMEWORK

In contrast to Uzbekistan, in Kazakhstan the implementation of the IWRM framework was somewhat possible as it was also formalized by the water code in 2003. Furthermore, for ten years the Kazakh government's involvement with international donors has been stronger in comparison to Uzbekistan's, having received the assistance of international donors in drafting of the law on WUAs and the Water Code (2003) as well as in implementing the project aimed at establishing the basin councils within the river basin organizations. Due to this involvement and their productive relations with the development agencies, since the 1990s, Kazakhstan has gained the reputation (shared with Kirghizstan) for being the most open Central Asian republic in terms of questioning its political-economic system and in undertaking institutional and economic reforms. Like Uzbekistan, Kazakhstan has received millions of dollars in support for these changes. Nevertheless, in the last years, except for the donors-based area project (such as the Makhtaral district in South-Kazakhstan province), the IWRM implementation and the IMT process have not been adequately supported and promoted (by the province and district-level state organizations) at the local level. Since several WUAs had already been established in 2003, these actors, particularly in the last years, have not provided any financial loans or know-how or technical support to the governing board of the WUAs which have been experiencing organizational issues. Despite the formalization of the IWRM and in conflict with the IMT, the district governments, instead of providing help to the WUAs in challenging conditions, continued to financially support the district water departments — some of whom were already in financial crisis and involved in dismantling processes. Support for the WUAs model, and hence to the IMT, has also been lacking from the water users in the last years. On the one hand, many water users have not financially supported the associations through feepayments, preferring the cheaper fees of the state organizations; on the other hand, they were not interested in strengthening, through a participatory approach, the newly independent organizations. Therefore, despite the IWRM formalization, an effective strengthen of the framework lacked both by the state organizations and part of the water users. Their lack of support legitimized the state's initiation of the recentralization process at the local level, which has been analyzed in depth above, through the action of the Republican State Enterprises towards the failed WUAs. It should be mentioned that this process of new state water control isalso financially possible due to the huge revenues coming from the oil and gas sectors. Therefore, ten years after the formalization of the IWRM, this recentralization process is totally in conflict with the framework's pillars and the IMT and surely indicates, from one standpoint, the crisis of the donors' promoted rationale, and from another standpoint the state's willingness to increase again — although with a significant difference from Uzbekistan, the water resource management control at the local level. Nevertheless, since this recentralization process was recently initiated and the future of the WUAs is uncertain, currently the whole structure of the local-level water sector seems uncertain and unforeseeable for the near future. Anyhow, this political turnaround towards a recentralization process marks a significant breaking point with the IWRM orientation of the last decade, despite its formalization.

## 8.3 THE CONTRIBUTION OF THE CENTRAL ASIAN LESSON TO THE INTERNATIONAL DEBATE: REDISCUSSING THE IWRM

Nowadays the whole implementation of the IWRM in Uzbekistan and Kazakhstan, as initially sponsored by the donors, is challenging, being hampered and reshaped by the political, social, and economic systems that are still somehow related to the inherited Soviet system — both in terms of the government and the civil society — that characterize the two countries. Both Uzbekistan and Kazakhstan gave a national interpretation to the IWRM implementation process, allowing their current systems to take advantage of the initiative and to legitimize their strategies. Therefore, since the promotion of the IWRM rationale in developing countries in the last decade has also included the willingness to introduce democratic principles, changes in hierarchies, and a reduction of the power of state bureaucratic systems, it is possible to affirm that the framework's rationale has not achieved these aims in either of two countries. In fact, these governments tried to implement only those parts of the pillars which did not question or change their hierarchical and bureaucratic structures. Hence the IWRM rationale was somehow not able to scratch the strong state and local water bureaucracies inherited from the Soviet Union, which have only slightly changed during the post-Soviet decades. Resuming the debate on the IWRM implementation discussed in depth in Chapter 1 — its definition, its aims, and whether it can fit within the local sociopolitical environments throughout the world — it can be questioned if this framework is the best strategy for improving water management in Central Asia. Though on the one hand it surely could lead to improvements — in particular in terms of environmental concerns, such as water conservation and reduction of the water use in relation with water economic mechanisms — on the other hand, the results have clearly shown that the framework, as presented on paper and promoted by the Global Water Partnership, is not wholly implementable. As discussed, it clashes with the sociopolitical and economic context characterizing the two countries and with the almost complete absence of involvement of the civil society in the decision-making processes — which is one of the fundamental requirements, according to the GWP, for implementing the IWRM. Therefore, the evidence emerging from the analysis of the IWRM implementation in Uzbekistan and Kazakhstan confirms what Biswas (2008) states and criticizes about the IWRM: according to him, it is impossible to implement a framework and its pillars worldwide without considering the different physical characteristics of the regions, the different importance of water for the environment, the quantity of water in the hydrological system, and, finally, the cultural, political, and economical milieu of the different countries.[462] As mentioned before, the main issues which hampered the whole implementation of the IWRM are basically related to the sociopolitical sphere and less to the natural or physical realms. Also the importance of the water resources for the environment and the state economy is significant: for instance, in an authoritarian state characterized by a wet environment it would be less strategic to promote the IMT in comparison with an arid or semi-arid country where water plays a strategic role in human life and economic production. According to Molle's analysis (2007), it seems clear that the IWRM's rationale is related to the interests and the political/economic environment of the place where it was created — that is, the Western world. The water professionals (raised and educated exclusively in developed Western countries, as mentioned in Chapter 1)

---

[462] BISWAS, A.K., 2008. "cit.".

who elaborated the framework did not pay enough attention to the potential issues and the challenges of implementing the IWRM framework in political and economic contexts different from their own. It was their hope that through the different implementation toolboxes, the framework would call into question and then be able to change the existing sociopolitical structures. These toolboxes sponsored by the GWP, as argued also by Biswas (2008), were quite general and unclear, including fascinating words and fuzzy practices, without providing effective guidance for the IWRM implementation. Nevertheless, in relation to the debate concerning this implementation process throughout the world, his position that the IWRM is not implementable worldwide and that the concept it is already in decline, appears a bit extreme. With a partial re-thinking and the strengthening of the GWP toolboxes, it would be possible to achieve a potential IWRM implementation in those countries and regions which already possess the prerequisites for a successful adaptation. The prerequisites should be a stable democratic political system, a mixed economic system already involving both the state and private actors (in particular regarding natural resources management), an active civil society and a structured perspective on environmental issues. For instance, in South Africa and in other countries of Southern Africa, as discussed by Van der Zaag (2004), these requirements have allowed a successful implementation of the IWRM in the last years.[463] However, it has been questioned whether this successful implementation referred only to a specious official international agreement or if it had led to tangible improvements in the livelihoods affected by water unavailability issues. The lesson learned from Central Asia has shown that in this region the above-mentioned requirements to support the process are almost absent; therefore, the framework as initially sponsored, is not wholly implementable without taking into account a national restructuring. Nevertheless, although several authors, mostly based in academia, have suggested different changes to the framework's pillars in the last years, as discussed in Chapter 1, according to Molle (2007), it would be very challenging for the GWP to question and re-think the IWRM because of the political and economic implications.[464] Millions of dollars, through loans and financial support, were invested and spent by the development banks and agencies to spread the IWRM, therefore, putting the framework into question would indicate a failure for all the involved water community professionals and financial supporters. It can be

---

[463] VAN DER ZAAG, P., 2004. "cit.".
[464] MOLLE, F., 2008. "cit.".

strongly criticized, as Merrey (2005) claims, that the donors which support the IWRM mostly focus on the introduction of economic principles with the aim of conserving water from an environmental perspective, instead of focusing on poverty alleviation and livelihood issues, such as drinking and irrigation water security.[465] Furthermore, in the last decade the GWP and the development agencies have tried to hide the political nature of the IWRM implementation process and its pillars, disguised as development aid, such as the "hydrographization" and the spread of the participatory approach. The implementation of these pillars require institutional changes and a rethinking of state rationale and procedures which cannot be separated from politics and related policies. Furthermore, as also widely debated by Molle (2008), the IWRM has a really strong, though somewhat hidden, ideological and political aim to change the existing political structures throughout the world, as evinced from their support of the rolling back of the state, to the privatization processes and the bank investments actions, and the introduction of monetary mechanisms. The evidence that emerges from the Uzbek and Kazakh case studies shows how the political nature and related implications of the IWRM implementation are strong, and also demonstrates how they aim to shape the governmental and social structures. However, it seems quite clear that in both countries this rationale, so far, has only been able to scratch the surface of a long-established cultural and political system which is quite the opposite of the one supported by the IWRM and the implementing agencies. Allan (2003) was one of the first scholars to argue that the IWRM and generally the current (since 2000s) water management discourse is a political process. As discussed in Chapter 1, Allan claims that the IWRM, to be implementable, requires the knowledge and the support of all the stakeholders (government, private actors, NGOs, and water users) involved in the process and without these requirements, implementation is extremely challenging or perhaps even impossible.[466] Hence, according to him, the KHWOE process, discussed in Chapter 1, is necessary to implement a reform path. What emerges from the two case studies is that in both countries the first point of the KHWOE, *knowing about* the proposed reforms is almost entirely lacking. In fact, except for the national and district donors projects areas where the development agencies tried to spread the reforms through seminars, most of the stakeholders at the local level, including the WUAs' staffs and the water users, have not been adequately in-

---

[465] MERREY, D.J., 2005. "cit.".
[466] ALLAN, T., 2001. "cit.".

formed about the water reforms undertaken at the national level. In some cases, in particular in the Middle Zeravshan valley, the water users did not even know they were part of a WUA, or about the shift from the state organization to the water users association. Rather than the donors, the Uzbek and Kazakh governmental authorities bear responsibility for this lack, in particular the Uzbek one, which did not want to spread the changes and involve the local level stakeholders, in order not to question its institutional stability and avoiding potential issues; this strategy clearly reflecting a continuation of a vertical top-down approach. Through this lack, an important part of the stakeholders were excluded from IWRM support and involvement in its implementation, and therefore, according to Allan, this dynamic already affected the process. The KHWOE's second point, to *want* (the reform), finally clearly allows the understanding of the evidence debated in this process. Excluding the involvement of the water users and the almost absent private actors and their potential influence, particularly in Uzbekistan, the two governmental authorities were able to shape the IWRM implementation process according to their specific *want*. Therefore, avoiding the complex KHWOE process, two different reinterpretations of the sponsored framework have emerged — both clashing with the initial one in various ways — according to the different state aims and related strategies. Hence in the final analysis it is possible to state, referring to the international debate, that in these two countries of the Central Asian region, the IWRM could be implementable, but according to a national way specific to each country — strongly shaped by local rationales — which remains quite far from the one sponsored by the initial donors. Therefore, a plural and multi-perspective Central Asian adaptation of the promoted IWRM has evidentially emerged, but with the prospects for the near future uncertain and complex.

# REFERENCES

ABDULLAEV, I. et Al. 2009. Agricultural Water Use and Trade in Uzbekistan: Situations and Potential Impacts of Market Liberalization, *Water Resource Development*, 25, 1.

ABDULLAEV, I., MOLLINGA,P., 2010. The Socio-Technical Aspects of Water Management: Emerging Trends at Grass Roots Level in Uzbekistan, *Water*, 2.

ADRIANOV, B.V. 1995. The Influence of Economic Development in the Aral Region and its Influence on the Environment, *Geojournal*, 35.1, pp.11-16.

AGNEW, J., 1994. The Territorial Trap: The Geographical Assumption of the International Relations Theory, *Review of International Political Economy*, 1,1.

ALLAN, T., 2001.*The Middle East Water Question- Hydropolitics and the Global Economy*, I.B. Tauris & CO Ltd.

ALLAN, T., 2003.IWRM/IWRAM: A new sanctioned discourse?, Occasional paper 50, SOAS/King's college University, London.

ALLOUCHE, J., 2005. *Water Nationalism: An Explanation of the Past and Present Conflicts in Central Asia, Middle East and the Indian Sub-continent?* PhD thesis, Universitè de Geneve.

ALLOUCHE, J. 2007. The governance of Central Asian waters: national interests versus  regional cooperation, (Central Asia at the crossroads), *Disarment Forum*, 4.

AMINOVA, M., ABDULLAEV, I., 2009. Water Management in State-Centered Environment: Water Governance analysis of Uzbekistan, *Sustainability*, 1.

BENJAMINOVIC, R., TERZINSKIY, P. 1975. *Irrigatzia Uzbekistana*, Nedatelstvo Fan, Uzbek CCP, Tashkent.

BENSIDOUN, S., 1979. *Samarkand et la Vallé du Zeravchan,* Anthropos, Paris.

BETHEMONT, J., 1999. *Les Grandes Fleuves. Entre Nature et Société,* Armand Colin, Paris.

BETHEMONT, J., FAGGI, P.P., ZOUNGRANA T.P., 2003. *La Vallé du Sourou (Burkina Faso): Genese d' un territorie hydraulique dans l'Afrique Soudano-Sahelienne,* l'Harmattan France.

BICHSEL, C., 2009. *Conflicts Transformation in Central Asia: Irrigation Disputes in the Fergana Valley,* Routledge, London & New York.

BICHSEL, C., 2011. Liquid Challenges: Contested Water in Central Asia, *Sustainable Development Law and Policy.*

BICHSEL, C., 2012. "The Drought Does Not Cause Fear": Irrigation History in Central Asia through James C. Scott's lenses, *Revue d' Etudes Comparatives Est-Ouest,* vol.43, n.1-2.

BISWAS, A.K., 2004. Integrated Water Resources Management: A Reassessment – A Water Forum Contribution, *Water International,* 29, 2.

BISWAS, A.K. 2008. Integrated Water Resources Management: Is it working?, *Water Resources Development,* vol. 24, n.1.

BISWAS, A.K., 2010: Management of International Waters: Opportunities or Constraints? *International Journal of Water Resources Development,* vol. 15, n.4.

BOCH, P., 2002. *Agrarian Reforms in Uzbekistan and in other Central Asian Countries,*working paper.n.49, Land Tenure Centre, University of Wisconsin, Madison.

BURGER, R., 1998. *Water Users Associations in Kazakhstan: an Institutional Analysis,* NIS PROJECT, Harvard Institute for International Development, Environment Discussion Paper n.45.

CARSON, R., 1965. *Silent Spring,* London, Penguin Books.

DJALALOV, A.A., 2001. *National Water Law of Uzbekistan: Its Coordination with International Water Law. Priorities and Problems. Line of Activities for Improvement,* ICWC Training Centre Seminar "International and National Water Law and Policy, September 24-29, 2001.

DOGAN, M., PELASSY, D., 1990.*How to Compare Nations: Strategies in Comparative Policies,* Chatman House Publisher.

DUKHOVNY, V.A., SOKOLOV, V.I., 2005. *Integrated Water Resources Management-Experience and Lessons Learned for Central Asia towards the Fourth World Water Forum.* Tashkent: SIC ICWC-GWP CACENA.

DUKHOVNY, V. et Al. 2008. IWRM implementation: Experience with Water Sector reforms in Central Asia, in *Central Asian Waters (Rahaman, Varis),* Water and development publications, Helsinki University of Technology.

DUKHOVNY, V. et Al. 2009. *Integrated Water Resources Management: Putting Good Theory into Real Practice. Central Asian Experience.*SIC ICWC, GWP CACENA, Tashkent.

DUKHOVNY, V., DE SCHUTTER, J., 2011.*Water in Central Asia: Past, Present and Future,* CRC press.

DURKHEIM, E. 1982.*The Rule of Sociological methods,* (W.D. Halls Trans), New York: the Free Press.

ELDEN, S., 2004. Missing the Point: Globalization, Deterritorialization and the Space of the World, *Trans Inst Br Geogr- Royal Geographical Society.*

ELDEN, S., 2010. Land, Terrain, Territory, *Progress in Human Geography,* 34:799.

EVERS, H.D., BENEDIKTER, S., 2009. Hydraulic bureaucracy in a modern hydraulic society: Strategic group formation in the Mekong Delta, Vietnam, *Water Alternatives,* 2(3).

FAGGI, P.P., 1986. Pour un géographie des grands projets d'irrigation dans les terres seches des pays sous-developpè: les impacts sur le milieu et leurs consequences, *Revue de Geographié du Lyon,* vol. 61 n.1.

FAGGI, P.P. et al., 1995. *Irrigazione, Stato e Territorio in Sudan; il gioco della posta in gioco*, Terra d'Africa,Milano, Unicopli.

FAGGI, P.P., et al., 2002. La valle del Sourou (Burkina Faso): per una geografia della territorializzazione in Africa, *Rivista Geografica Italiana,* vol. 109, n.2.

FEDCHENKO, F. 1870. "Topographical Sketch of the Zarafshan Valley, *Journal of the Royal Geographical Society of London,* vol.40, pp. 448-461.

GHAZOUANI, W. et al, 2012.*Water Users Association in the NEN region_IFAD interventions and overall dynamics,* IWMI, IFAD.

GRANIT, J. Et al. 2010.*Regional Water intelligence: report on Central Asia,* UNDP-SIWI, paper n. 15.

G.W.P., TAC, 1998. *IWRM- At a glance,* GWP document, Stockholm.

G.W.P., TAC 4, 2000. *Integrated Water Resources Management,* Global Water Partnership, Technical Advisory Commitee Paper, Stockholm, Sweden.

G.W.P., 2006.*What is GWP?,*Stockholm, Sweden.

G.W.P. and UCC-Water, 2006.*The Republic of Uzbekistan, National Report- Within the framework of UNEP support for achieving the Johannesburg Plan of implementation target of "Integrated Water Resources Management and Efficiency Plans by 2005, with support to developing countries".* Tashkent.

G.W.P and UCC-Water, 2006.*"Road Map" Planned Steps towards Realization of the Integrated Water Resources Management Principles and Rationale of the Essential Activities in the Republic of Uzbekistan.* Tashkent.

GLOBAL WATER PARTNERSHIP, 2009.*A Handbook for Integrated Water Resources Management in Basins,* Elanders, Sweden.

GOVERNMENT OF THE REPUBLIC OF KAZAKHSTAN /UNDP, 2004.*National Integrated Water Resources Management and Efficiency Plan in Kazakhstan,* Project document.

GRAEFE, O., 2011. River basins as new environmental regions? The depolitization of water management, *Procedia Social and Behavioral Sciences,* 14.

GUMBO, B., VAN DER ZAAG, P., 2001. *Principles of Integrated Water Resources Management (IWRM),* Global Water Partnership (GWP) Southern Africa, Southern Africa Youth Forum, 24-25 September, Harare, Zimbabwe.

GUNCHINMAA T. ,YAKUBOV, M. 2009. Institutions and Transition: does a better institutional environment make the water users' associations more effective in Central Asia?, *Water Policy,* 1, 22.

HORNIDGE, A,K., et Al., 2011. Reconceptualizing Water Management in Khorezm, *Natural Resources Forum.*

HUNT, R., 1989. Appropriate Social Organization? Water Users Associations in Bureaucratic Canal Irrigation System, *Spring,* vol.48.

ICG, 2002.*Central Asia: Water and Conflicts,* ICG Asia report, n.34.

IWMI, 2012.*Review on WUAs' development in Uzbekistan,* unpublished report.

ILKHAMOV, A., 1998. Shirkat, Dekhqon farmers and others: Farm restructuring in Uzbekistan, *Central Asian Survey,* 17:4.

ISAMIDDINOV, M., 2002.*The Irrigation Development of the Samarkandian Sogd in the Ancient Times*, Izadeltstvo, Tashkent.

JONKER, L. (not Ment.). Integrated Water Resources Management: the theory-praxis-nexus, IWRM programm, University of the Western Cape, South Africa, Working Paper.

JOSEPHSON, P.R., 1995. "The Project of the Century" in the Soviet Union: Large-scale technologies from Lenin to Gorbachev, *Technologies and Culture,* 36:3.

JOZAN, R. 2008. "Etat délinquant" ou modèle deviant? Retour sur le non respect de traite international de partage de la ressource en eau du Syr-Darja, *Metropolis/Flux,* 1, n. 71.

KANDIYOTI, D., 2002. *Agrarian Reforms, Gender and Land Rights,* Social Policy and Development Programme Paper n.11, United Nations Research Institute for Social Development.

KARAR, E., 2008. *Integrated Water Resources Management (IWRM): Lessons for Implementation in the Developing Countries,* presented at the International Conference on IWRM, Cape Town, South-Africa.

KEMAN, H., 1993. *Comparative Policies: new Directions in Theory and Methods,* Amsterdam, VU Press.

LANDMAN, T. 2000. *Issues and Methods in Comparative Policies, an introduction,* Routledge, London and New York.

LENDERTSEE, K., et al., 2008. *IWRM and the environment: a view on their interactions and examples where IWRM led to better environmental management in the developing countries,* presented at the International Conference on IWRM, Cape Town, South-Africa.

LERMAN, Z. & STANCHIN, I. 2004. Institutional Change in Turkmenistan' Agriculture: Impacts on Productivity and Rural Incomes, *Eurasian Geography and Economics,* 45, n.1.

LEWIS, R., 1962. The Irrigation Potential of Soviet Central Asia, *Annals of Association of American Geographers,* vol. 52, n.2.

LEWIS, R. 1966. Early Irrigation in West Turkestan, *Annals of the Association of the American Geographers,* vol. 56, n.3.

LEWIS, R. 1992. *The Geographical Perspectives on Soviet Central Asia,* Routledge, London.

MAHONEY, J., RUESCHEMEYER, D., 2003. *Comparative Historical Analysis in the Social Sciences,* Cambridge: Cambridge University Press.

MAINGUET, M. 1995. *L' Homme et la Secheresse,* Masson, Paris.

MATLEY, I,. 1970. The Golodnaya Steppe: a Russian Irrigation Venture in Central Asia, *Geographical Review,* vol. 60, n.3.

MERREY, D. 1996. *Institutional Design principles for Accountability in Large Irrigation Systems,* International Irrigation Management Institute (IIMI), research paper 8.

MERREY, D.J., et al., 2005. Integrating Livelihoods into Integrated Water Resources Management: taking the integration paradigm to its logical next step for developing countries, *Regional Environmental Change,* 5.

MICKLIN, P. 2007. The Aral Sea Disaster, *Annual Review of Earth and Planetary Sciences,* 35.

MOLLE, F., 2008. Nirvana Concept, Narratives and Policy Models: Insights from the Water Sector, *Water Alternatives,* 1 (1).

MOLLE, F., 2009. River-basin Planning and Management: the Social Life of A Concept, *Geoforum,* 40.

MOLLE, F., MOLLINGA, P., WESTER, P., 2009. Hydraulic Bureaucracies and the Hydraulic Mission: Flows of Water, Flows of Power, *Water Alternatives,* 2(3).

MOLLINGA, P., 2007. *Water Policy-Water Politics: Social Engineering and Strategic Action in Water sector Reforms,* ZEF Working Paper series, n.19. University of Bonn.

MOLLINGA, P., 2008. Water, Politics and Development: Framing a Political Sociology of Water Resources Management, *Water Alternatives*, 1(1).

MOLLINGA, P., 2010. Boundary Concepts for Interdisciplinary analysis of Irrigation Water Management in South-Asia, ZEF Working Paper Series 64, University of Bonn.

MOLLINGA, P., GONDHALEKAR, D., 2012. Theorizing Structured Diversity. An approach to comparative research in water resources management, ICCWaDS, working paper n.1.

MOTTMACDONALD-DFID, 2003.*Privatization / Transfer of Irrigation Management in Central Asia,* final Report.

MOSSE, D., 2008. Epilogue-The Cultural Politics of Water- A Comparative Perspective, *Journal of Southern African Studies,* 34:4.

O'HARA, S., HANNAN, T., 1999.Irrigation and Water management in Turkmenistan: Past Systems, Present Problems and Future Scenarios, *Europe-Asia Studies,* vol. 51, n.1.

OLSSON, O., et al., 2010. Identification of the effective water availability from streamflows in the Zeravshan river basin, Central Asia, *Journal of Hydrology,* 390, 170-177.

OSTROM, E.. 1990. *Governing the Commons: The Evolution of Institutions for Collective Action.* Cambridge University Press.

OYEN, E. (ed.), 1990. *Comparative Methodology: Theory and Practice in International Social Research,* International Sociological Association, Sage publications.

PIASTRA, S., 2012. *Land Reclamation: Geo-Historical Issues in a Global Perspective,* Proceedings of the International Conference held at the University of Bologna, Patron ed.

POMFRET, R., 2007. Rebuilding Kazakhstan's Agriculture, *Central Asia and Caucasus Analyst.*

RAFFESTIN, C., 1981. *Pour une Geographié du Pouvoir,* Paris, ed. LI TEC.

RAGIN, C.C., 1987. *The Comparative Method: Moving beyond the qualitative and quantitative strategies,* University of California Press.

RAGIN, C., 1991. *Issues and Alternatives in Comparative Social Research,* E.J. Brill.

RAGIN C.C., 1997. Turning the Tables: How case-oriented research challenges variable-oriented research, *Comparative Social Research,* 16.

RAHAMAN, M., VARIS, O., 2005.Integrated Water Resources Management: Evolution, Prospects and Future Challenges, *Sustainability, Science, Practice and Policies,* vol.1, n.1.

RAHAMAN, R. 2012. Water Wars in 21$^{st}$ century along International River Basins: speculation or reality? *Int. J. Sustainable Society,* vol. 4, n.1/2.

RAMAZANOV, A.M., 2001. *National Water Law of Kazakhstan:Its Coordination with International Water Law. Priorities and Problems. Line of Activities for Improvement,* ICWC Training Centre Seminar "International and National Water Law and Policy, September 24-29, 2001.

RICKMERS, R. 1913. *The Duab of Turkestan: A Physiographic sketch and account of some travels,* Cambridge University Press, Cambridge.

SALMAN, M.A (1997) *The Legal Framework for Water Users' Associations-A comparative study.*World Bank Technical Paper n.360.

SEHRING, J., 2007. *The politics of Water Institutional Reform in Neo-Patrimonial States: a comparative analysis of Kirghizstan and Tajikistan,* PhD thesis, Fern Universitet in Hagen.

SEHRING, J, 2007. Irrigation Reforms in Kirghizstan and Tajikistan, *Irrigation Drainage Systems,* 21.

SMITH, D. 1995. Environmental Security and Shared Water Resources in Post-Soviet Central Asia, *Post-Soviet geography,* 36, 6.

SNELLEN, W.B., A. SCHREVEL, 2004. *IWRM: for sustainable use of water: 50 years of international experience with the concept of Integrated Water Management,* Background document to the FAO, Netherlands Conference on Water for Food and Ecosystem.

SOLANES, M., GONZALEZ VILLAREAL, F., (GWP-Tac), 1999.*The Dublin Principles for Water as Reflected in a Comparative Assessment of Institutional and Legal Arrangements for Integrated Water Resources Management,* GWP TAC Background Paper n.3.

SPOOR, M., KRUTOV, A, 2002. The Power of Water in a divided Central Asia.

SPOOR, M., 2006. Uzbekistan' s Agrarian Transition, in: Chandra Suresh Babu and Djalalov S. (eds.) *Policy Reform and Agricultural Development in Central Asia,* Boston, Springler.

STRIDE, S., et al., 2009. Canals versus Horses: Political Power in the oasis of Samarkand, *World Archeology.* 41 (1).

SUSLOV, S.P. 1961. *Physical Geography of Asiatic Russia,* WH Freeman, San Francisco.

THURMAN M. 2002: Irrigation and Poverty in Central Asia. A Field Assessment.http://www.esd.worldbank.org/bnwpp/documents/7/IrrigandPoverty InCAvers2.pdf.

TILLY, C., 1984. *Big structures, Large Processes, Huge Comparisons,* New York, Russel Sage Foundation.

TOLSTOV, S.P., 1948. *Drevnii Khorezm,* Moskva, Izdanie MGU.

TREVISANI, T., 2007. After the Kolchoz: rural elites in competition, *Central Asian Survey,* 26:1.

TURCO, A., 1988. *Verso una teoria geografica della complessità,* Unicopli, Milano.

UNDP, 2007.*Integrated Water Resource Management and Water Efficiency in the Republic of Kazakhstan, 2008-2025,* Progress Report, UNDP PROJECT #35289.

UNDP-GOVERNMENT OF NORWAY, 2005. *Kazakhstan National Integrated Water Resource Management and Efficency Plan,* draft of a project document.

VAN DER ZAAG, P., 2001. *Principles of Integrated Water Resources Management,* Waternet module IWRM 0.1, 1st draft, HIE-Delft and Department of Civil Engineering, University of Zimbabwe.

VAN DER ZAAG, P., (2005). IWRM: relevant concept or irrelevant buzz-word? A capacity building and research agenda for Southern Africa, *Physics and Chemistry of the Earth,* 30.

WEGERICH, K., 2000. *Water Users Associations in Uzbekistan and Kirghizstan:Study on Conditions for Sustainable Development,* Occasional Paper n.32, Water Issue study group,SOAS, IWMI.

WEGERICH, K. 2005. *Institutional changes in Water Mangement at local and Provincial level in Uzbekistan,* Bern et. Al.

WEGERICH, K., 2006. *"Handing over the sunset"- External factors influencing the establishment of Water Users Associations in Uzbekistan: Evidences from Khorezm Province,* Phd thesis, Humboldt Universitet zu Berlin.

WEGERICH, K., 2008. Hydro-Hegemony in the Amu-Darja Basin, *Water Policy,* 26(2).

WEGERICH, K., 2008. Blueprints for water users associations' accountability versus local reality: evidence from South Kazakhstan, *Water International,* vol.33, n.1

WEGERICH, K. 2011. Water Resources in Central Asia: regional stability or patchy make-up?*Central Asian Survey,* 30:2.

WEGERICH, K. et Al., 2012. Meso-Level Cooperation on Transboundary Tributaries and Infrastructures in the Fergana Valley, *International Journal of Water Resources Development,* 28:3.

WEGERICH et al, 2012. Is It Possible to Shift to Hydrological Boundaries? The Fergana Valley Meshed System, *International Journal of Water Resource Development,* vol.8, n.3.

WESCOAT, J.L., 2003. Water Resources. In: Gaile G.L. and C.J. Willmott (eds.) 2003. *Geography in America at the Dawn of the 21ˢᵗ Century,* Oxford University Press.

WIMMER, A., DE SOYSA, I., WAGNER, C., 2003. *Political Science Tools for Assessing Feasibility of Reforms* (ZEF Discussion Papers on Development Policies, n. 63) Bonn.

WITTFOGEL, K., 1957. *Oriental Despotism: a Comparative Study of Total Power*, Yale University Press, New Haven.

WORLD BANK, 2011. *Implementation Status and Results, Uzbekistan, Rural Enterprise Support Project II,* Report n. ISR6177.

WORLD BANK, 2012. *World Bank Development Indicators"*.

WORLD BANK, 2012. *Proposed Project Restructuring of Second Rural Enterprise Support Project,* Restructuring Paper from the WB to the Republic of Uzbekistan.

YALCIN, R., MOLLINGA, P. 2007. *Water Users Associations in Uzbekistan.The Introduction of a New Institutional Arrangement in Local Water Management.Amu-Darja case study, Uzbekistan,* NeWater project, University of Bonn.

YALCIN, R., MOLLINGA, P., 2007. *Institutional Transformations in Uzbekistan's Agricultural and Water Resources Administration: The Creation of a New Bureaucracy,* ZEF Working Paper Series 22, University of Bonn.

YAKUBOV, UL-HASSAN, 2007. Mainstreaming rural poor in water resources management: preliminary lessons of a bottom-up WUA development approach in Central Asia, *Irrigation and Drainage,* 56.

YAKUBOV, M., 2012. A Programm Theory Approach in Measuring Impacts of Irrigation Management Transfer Interventions: The Case of Central Asia, *International Journal of Water Resource Development,*28:3.

ZIGANSHINA, D. (no date). International Water Law in Central Asia: Commitments, Compliance and Beyond, *The Journal of Water Law,* 20.

ZIMINA, L., 2003. Developing Water Management in SouthKazakhstan, in *Drop by Drop: Water Management in the Southern Caucasus and Central Asia,* edited by S. O'hara, Local Government and Public Service Reform Initiative, LGI Fellowship Series, Open Society Institute.

ZINZANI, A., 2011. Tra Irriguo e Seccagno: le Trasformazioni Recenti nella Media Valle dello Zeravshan, Uzbekistan, *Rivista Geografica Italiana,* 118.

ZINZANI, A., 2014. Irrigation Management Transfer and the Dynamics of WUAs: Evidence from South-Kazakhstan Province, *Environmental Earth Sciences,* DOI 10.1007/s12665-014-3209-6.

# IMAGES FROM CENTRAL ASIA

*IMG. 1. The peaks and the glaciers of the Zeravshan range ( Western Alay mountains), chateaux d'eau of the Zeravshan river, Tajikistan (Sogd province)* [467]
.

*IMG.2. The Zeravshan river's bed (braided) before entering in Samarkand, Uzbekistan (Samarkand province).*

---

[467] IMG.1-2 courtesy of the Archeological Expedition of the University of Bologna; IMG. 3-15 by the author

*IMG 3. A section of the 1ˢᵗ May dam, where the Zeravshan canals system arises, Uzbekistan (Samarkand province).*

*IMG. 4 Chakylayan foothill in the Urgut district; the spring snow melt contributes to the irrigation of this area, included in Urgut WUA, Uzbekistan (Samarkand province).*

*IMG 5. The Yangyarik canal irrigates the central-southern section of the Urgut district, Uzbekistan (Samarkand province).*

*IMG 6. The border between an hydraulic territory (Karakalpakstan irrigated area) and the Kizilkum desert, Uzbekistan (Karakalpakstan Autonomous Republic).*

IMG 7. *Village in the Khorezm irrigated area, supplied by the Amu Darja's streams, Uzbekistan (Khorezm province).*

IMG. 8: *Cotton fields irrigation in the territory of Pastdargom WUA, Uzbekistan (Samarkand province).*

*IMG 9. The Ulus irrigated area, -included in Nurabad WUA and supplied by the Eski-Anghor canal-, surrounded by the steppes, Uzbekistan (Samarkand province).*

*IMG 10: Talaiski-Alatau range (Western Tian-Shan mountains) and related foothill area, chateaux d'eau of the Arys river and its tributaries, Kazakhstan (South-Kazakhstan province).*

*IMG 11: Talaiski-Alatau glaciers, Kazakhstan, (South-Kazakhstan province).*

*IMG. 12: The upstream section of the Arys valley, included in the Tyulkibas WUA, Kazakhstan (South-Kazakhstan province).*

IMG. 13. *Irrigated schemes surrounded by the steppes in the downstream section of the Arys valley included in Otrar WUA, Kazakhstan (South-Kazakhstan province).*

IMG. 14. *The Arys river in the Ordabasy WUA's territory, close to the intake of the Arys-Turkestan canal, Kazakhstan (South-Kazakhstan province).*

*IMG. 14: View of the Koksarai dam, built in 2010 to storage the winter flow of the Syr-Darja, Kazakhstan (South-Kazakhstan province).*

*IMG. 15: The author with the director of the Karaspan WUA, Ordabasy district, (South-Kazakhstan province).*